COLLIMATORS FOR THERMAL NEUTRON RADIOGRAPHY

AN OVERVIEW

COLLIMATORS FOR THERMAL NEUTRON RADIOGRAPHY
AN OVERVIEW

ABSTRACT

The purpose of this survey is to review the design and construction of collimators used for thermal neutron radiography. To this end handbooks and general publications in the field of neutron radiography have been reviewed. Conference papers related to non-destructive testing and neutron radiography were also considered. As the resultant information was rather sparse and incomplete, the author attempted to extract information concerning collimators from other sources, particularly the numerous papers describing neutron radiography facilities in different countries. For practical reasons it was not possible to review all the papers available on the subject. A concise overview is presented concerning the basic data concerning the design and construction of collimators used for thermal neutron radiography.

Commission of the European Communities

Joint Research Centre, Petten Establishment

Neutron Radiography Working Group (NRWG)

COLLIMATORS FOR THERMAL NEUTRON RADIOGRAPHY

AN OVERVIEW

Compiled by

J. C. DOMANUS

Risø National Laboratory, Denmark

and

Edited by

J. F. W. MARKGRAF

*Commission of the European Communities,
Joint Research Centre, Petten Establishment,
The Netherlands*

D. REIDEL PUBLISHING COMPANY

A MEMBER OF THE KLUWER ACADEMIC PUBLISHERS GROUP

DORDRECHT / BOSTON / LANCASTER / TOKYO

Library of Congress Cataloging in Publication Data

Collimators for thermal neutron radiography: an overview / compiled by J.C. Domanus
and edited by J.F.W. Markgraf.
p. cm.
Bibliography: p.
ISBN 90-277-2568-3
1. Neutron radiography. 2. Collimators (Optical instrument)—Design and construction. I. Domanus, J.C. II. Markgraf, J.F.W.
QC793.5.N4628C63 1987
539.7′213—dc 19

87-17630
CIP

Design frontcover: J. Wells, ISPRA
Lay-out: Reproduction service J.R.C. PETTEN

Publication arrangements by
Commission of the European Communities
Directorate-General Telecommunications, Information Industries and Innovation, Luxembourg.

EUR 10859 EN
© 1987 ECSC, EEC, EAEC, Brussels and Luxembourg

LEGAL NOTICE
Neither the Commission of the European Communities nor any person acting on behalf of the Commission is responsible for the use which might be made of the following information.

Published by D. Reidel Publishing Company,
P.O. Box 17, 3300 AA Dordrecht, Holland.

Sold and distributed in the U.S.A. and Canada
by Kluwer Academic Publishers,
101 Philip Drive, Assinippi Park, Norwell, MA 02061, U.S.A.

In all other countries, sold and distributed
by Kluwer Academic Publishers Group,
P.O. Box 322, 3300 AH Dordrecht, Holland.

All Rights Reserved
No part of the material protected by this copyright notice may be reproduced or
utilized in any form or by any means, electronic or mechanical,
including photocopying, recording or by any information storage and
retrieval system, without written permission from the copyright owner.
Printed in The Netherlands

PREFACE

One of the aims of the Neutron Radiography Working Group (NRWG) - constituted under the auspices of the Commission of the European Communities,*) is the collection and dissemination of information on different aspects of neutron radiography. Since its constitution in 1979 the NRWG has contributed to the publication of such books as the "Neutron Radiography Handbook" (EUR 7622e, 1981), the "Neutron Radiography Proceedings of the First World Conference", San Diego, California, U.S.A., December 7-10, 1981 (EUR 8296 EN, 1983) and "Reference Neutron Radiographs of Nuclear Reactor Fuel" (EUR 8916 EN EP). Presently, the second, revised and enlarged edition of the NR handbook is in preparation. It will be issued as "Practical Neutron Radiography" in the same series of publications.

Whereas both editions of the NR handbook intend to give general information about the subject, some detailed information about particular problems deemed important for NR will be treated in special technical reports. Such will be the case for the subject of "Neutron Radiography on Nitrocellulose Film" (to be published soon) and this current report on "Collimators for Thermal Neutron Radiography". This specialised publication provides the core of information needed for the design and construction of any NR facility.
As it is intended to treat the theoretical problems connected with the choice and design of a collimator in "Practical Neutron Radiography", this report confined itself to practical data concerning collimator design and construction. It is hoped that this information will be useful for designing new or redesigning existent NR facilities.

J.F.W. Markgraf
JRC Petten
HFR Division
(Editor)

---

*) (represented by the Joint Research Centre, Petten Establishment, Petten, The Netherlands)

CONTENTS

LIST OF ILLUSTRATIONS 1

LIST OF TABLES 3

1. INTRODUCTION 5
2. HANDBOOKS AND GENERAL PUBLICATIONS 6
3. CONFERENCES ON NDT AND NR 13
4. DESCRIPTION OF NR FACILITIES 24
    4.1. Australia 24
    4.2. Belgium 24
    4.3. Brazil 26
    4.4. Canada 27
    4.5. Czechoslovakia 28
    4.6. Denmark 28
    4.7. France 29
    4.8. Germany 42
         FRG
         GDR
    4.9. India 49
    4.10. Indonesia 51
    4.11. Iran 51
    4.12. Iraq 51
    4.13. Israel 53
    4.14. Italy 55
    4.15. Japan 57
    4.16. Mexico 58
    4.17. The Netherlands 59
    4.18. Sweden 65
    4.19. United Kingdom 67
    4.20. U.S.A. 69
    4.21. U.S.S.R. 82
    4.22. Yugoslavia 83
    4.23. Final remarks 83
5. SUMMARY OF DESIGN DATA 85
    5.1. Geometric shape of the collimator 85
    5.2. Walls and their lining 87
    5.3. Filling of the collimator 91
    5.4. Shutters and diaphragms 91
    5.5. Gamma-ray filters 92
    5.6. Seals for in-pool collimators 93

REFERENCES 94

LIST OF ILLUSTRATIONS

Fig. 1. Accelerator-based thermal neutron radiography system 7
Fig. 2. Lay-out of the neutron radiographic facilities at the Battelle research reactor 8
Fig. 3. Divergent, conical collimator 10
Fig. 4. Principle of in-pool radiography 10
Fig. 5. NR facility of the Triton reactor 11
Fig. 6. Collimators of the mini-reactor. a-axial, b-tangential 12
Fig. 7. NR apparatus at the BRR 13
Fig. 8. Experimental arrangement for NR Toshiba Training Reactor 14
Fig. 9. Collimating system 15
Fig. 10. Watertight joints 16
Fig. 11. Collimating system for subthermal cold neutron radiography 18
Fig. 12. Lay-out of the NR facility at BR1 20
Fig. 13. Neutron radiography bench 21
Fig. 14. NR facility at Studsvik R2 reactor 22
Fig. 15. General and detailed lay-out of the BR2 facility 25
Fig. 16. Divergent collimator at CDTN 26
Fig. 17. Submerged NR facility at IEA-R1 27
Fig. 18. Arrangement for neutron radiography in the NRX reactor 28
Fig. 19. Risø double beam neutron radiography facility 28
Fig. 20. Underwater neutron radiography 30
Fig. 21. NR facility at the lateral channel of the Triton reactor 31
Fig. 22. Collimator of the lateral channel NR facility of the Triton reactor 31
Fig. 23. Collimator of the Osiris facility 32
Fig. 24. Evolution of the Isis collimating system 33
Fig. 25. Collimators of the Isis and Osiris facilities at Saclay 34
Fig. 26. The "dry" NR facility at Isis reactor at Saclay 35
Fig. 27. NR facility at the Orphée reactor at Saclay 36
Fig. 28. Conical collimator 37
Fig. 29. NR facility at the Melusine reactor at Grenoble 38
Fig. 30. Lay-out of the mini-reactor for NR 39
Fig. 31. Cutaway view of Mirene 40
Fig. 32. NR facilities at the Rapsodie and Phenix reactors at Cadarache 40
Fig. 33. Reactor installation scheme (LDAC and CEI) 41
Fig. 34. Neutron radiography facility at FR2 in Karlsruhe 42
Fig. 35. Divergent collimator at FR2 43
Fig. 36. Details of collimator used at FR2 44
Fig. 37. NR-1 facility of the FRJ-1 reactor at Jülich 46
Fig. 38. GENRA I NR facilities at the FRG-1 reactor 47
Fig. 39. GENRA I NR facility at the FRG-1 reactor 47
Fig. 40. GENRA II NR facility of the FRG-2 reactor 48
Fig. 41. NR facility at the RFR reactor at Rossendorf 49
Fig. 42. General layout and beam tube details of the NR facility of Kalpakkam 50

Fig. 43.	Collimator of the Triga Mark II reactor at Bandung	51
Fig. 44.	NR facility at the IRT-2000 reactor in Baghdad	52
Fig. 45.	Cross-section of the beam tube at the Soreq reactor	53
Fig. 46.	Construction of the neutron inlet collimator at Soreq	54
Fig. 47.	Pinhole arrangement of the collimator at Soreq	54
Fig. 48.	NR facility at Casaccia	56
Fig. 49.	Exposure device for simultaneous neutron and gamma radiography using a ^{252}Cf source	57
Fig. 50.	NR facility at the KUR reactor at Kyoto University	58
Fig. 51.	NR facility attached to E-2 experimental hole of the KUR	58
Fig. 52.	Transversal view of the TW1 collimation system	59
Fig. 53.	Collimator System of the LFR at ECN	60
Fig. 54.	Underwater NR installation in the pool side facility of the HFR, Petten	61
Fig. 55.	Collimator of the HFR, Petten	62
Fig. 56.	Collimator in the beamtube HB8 of the HFR reactor	63
Fig. 57.	Details of inlet diaphragm of collimator from fig. 56	63
Fig. 58.	New underwater neutron radiography installation at the HFR	64
Fig. 59.	Modification of the collimation system at the HB-8 dry NR facility	64
Fig. 60.	NR facility at the R2-0 reactor at Studsvik	66
Fig. 61.	Details of the NR facility shown in fig. 60	67
Fig. 62.	Thermal neutron collimator of the Dido reactor at Harwell	68
Fig. 63.	Lay-out of NR facility at the Herald reactor at Aldermaston	68
Fig. 64.	NR facility at the Viper pulsed reactor at Aldermaston	69
Fig. 65.	Treat reactor at ANL, Idaho Falls	70
Fig. 66.	NRAD facility of the ANL, Idaho Falls	71
Fig. 67.	NR test arrangement at Atomics International	72
Fig. 68.	NR facility at L-88 reactor of Atomics International	73
Fig. 69.	Beam port of the Triga Mark I NR facility	73
Fig. 70.	GE Nuclear Test Reactor NR facility	74
Fig. 71.	Collimator for NR at HEDL Triga reactor	75
Fig. 72.	Plan and elevation view of the NERF facility at the Livermore LPTR reactor	77
Fig. 73.	Schematic diagram of the NBS TNRF facility	78
Fig. 74.	Second ACPR NR tube of SNLA	79
Fig. 75.	Collimator of the NR facility of University of Missouri	80
Fig. 76.	NR facility in the Oak Ridge research reactor	81
Fig. 77.	NR facility at Texas A & M University	82
Fig. 78.	NR facility at Triga Mark II reactor in Ljubljana	83
Fig. 79.	Total cross section curves for B, Cd, Dy and In	87

LIST OF TABLES

Table 1. French NR facilities, constructed between 1966 and 1970...................29

Table 2. Shape of the collimators 86

Table 3. Walls of the collimators...88

Table 4. Materials used for lining 90

Table 5. Diaphragms and shutters 92

1. INTRODUCTION

As J.P. Barton quite rightly put it: "In neutron radiography, the design of the neutron collimator can be of equal importance to the choice of the neutron source and the method of neutron photography" /1/. In view of this a literature search was made to collect information on the design of collimators used in thermal neutron radiography. In the same paper Barton concludes that the best solution is to use a divergent-beam collimator.
The conclusion remains valid and this design is now the most commonly used at all neutron radiography facilities. The present review is limited to these types of collimators.
In recognizing the importance of the collimator design for neutron radiography, it could be expected that much attention would be devoted to the problem in handbooks and general publications on neutron radiography. Unfortunately, this is not the case. Even in papers describing particular neutron radiographic facilities the necessary information can not always be found.

This review was made in anticipation of the needs of all those who might design a new neutron radiographic facility in the future or would like to re-design existing equipment. The aim was to extract all useful information from the available literature about collimators for thermal neutron radiography.

Another aspect of this search was to collect practical information on the subject for inclusion in the second edition of the "Neutron Radiography Handbook" /2/, now in preparation by the Euratom Neutron Radiography Working Group (NRWG) (to be published as "Practical Neutron Radiography").

In the review given below data on collimators contained in handbooks and publications of general character on neutron radiography (NR) is first described. Thereafter, information from papers describing particular NR facilities is analyzed from the point of view of collimator design.

In describing the collimator design special attention is paid to its geometric shape and dimensions as well as to the materials used in its construction. Devices applied at the inlet and outlet of the collimator are also taken into account.

The first book on neutron radiography was published by H. Berger /3/ as early as 1965; during the following 20 years several other books and a large number of papers have been published on the subject. It is not possible to review them all in the present report, but it is hoped that the most important publications have been considered.
After reviewing the available data a summary is given of the dimensions and materials used for collimators.

In the review given below collimators are described which are used in different NR facilities, without regard to the current use of those facilities. The purpose of this survey is to review the design and construction of details of collimators that are currently in use or have been extensively used in the past.

2. HANDBOOKS AND GENERAL PUBLICATIONS

The handbooks and general publications on NR are reviewed here in chronological order.
In the early years of NR applications the opinion prevailed that parallel beams of neutrons were needed to obtain adequate neutron beam flux at the object to be radiographed. Berger /3/ has written in his basic book on NR that "the present state of art is such that parallel beams are definitely preferred".
Consequently in Bergers book /3/ one cannot find any recommendations about the design of a divergent collimator.

This statement, made in 1965, was no longer valid during the consequent development of NR, and to-day divergent beam collimators are quite commonly used; Barton /1/ has rightly advocated their use in 1967.

In /3/ some mention is made of shutters used in NR facilities. They can vary from simple boral sheets, to lithium fluoride, to masive Masonite, boral and lead combinations. According to Berger /3/ "for a relatively clean thermal neutron beam, a simple boral sheet can work very well".

The first conference about neutron radiography: "Radiography with neutrons" was held on 10-12.5.1973 at the University of Birmingham. Papers presented at this conference were published in 1975 by the British Nuclear Energy Society /4/.

The next NR handbook appeared 10 years later. Although the book on principles of neutron radiography /5/ written by Tyufyakov and Shtan in 1975 contains a special chapter on practical problems of NR one cannot find there any general recommendations regarding collimator design.
The authors describe a pilot plant used by them for NR (heavy water reactor, thermal neutron beam) in which the original cylindrical collimator was replaced "by a collimator in the form of a truncated cone with a diameter at the outlet of 300 mm. The internal wall of the collimator was made of steel 4 mm thick covered with a layer of paraffin and borax mixture 200 mm thick and then with a layer of cast iron, 150 mm thick". As this NR installation is of a swimming pool type, special shielding of the collimator is not required because the surrounding water performs the shielding.

At the inlet of the collimator described in /5/ replaceable diaphragms are provided, but no information is given either about their dimensions, shape or material. To reduce the gamma dose rate to approximately 1 rad/h, a 80 mm thick lead filter is placed in the collimator.

As the NR facility described in /5/ is a swimming pool type, it is comparatively easy to cut off the radiation beam by filling the collimator or part of it with water from the swimming pool or from a separate container. As the authors of /5/ quite rightly stress "the presence of even traces of water anywhere on the path of the neutron beam from the input of the diaphragm of the collimator to the neutron image detector is highly undesirable. Therefore all the detachable joints of the collimator and the radiography chamber placed right in the swimming pool must be airtight and the plants must have devices for removing traces of water from spaces accessible to it. This problem is often solved by blowing dry air through the spaces".

The next publication on NR of a general character in which it could be expected to find the relevant information on collimators is the ASTM Special Technical Publication 586 /6/ from 1976.

In the first paper of this publication /6/ J. P. Barton repeats his previous statement (quoted above from /1/) that "neutron beam collimation can be one of the most important variables in the technique... For the basic method where thermal neutrons are used it is important to have a fairly long and well designed collimator... In practice, it is the divergent beam type that is now nearly always used" /7/.

The next paper in the ASTM publication /6/ describes a collimator in a 3.6 MV van de Graaf accelerator NR facility /8/. For a distance of 1.27 m from the 38 mm diameter collimator port the L/D ratio is 33. The collimator has a cadmium neutron shield, 0.76 mm thick. As a means of helping to suppress the secondary source of neutrons a 100 mm thick wall of borated polyethylene is interposed between the moderator assembly and the exposure plane (see fig. 1). A port in this neutron shield aligns with the moderator extraction channel to allow free passage of the radiography beam. A further ported shield, composed of 200 mm of lead is installed to help suppress gamma rays.

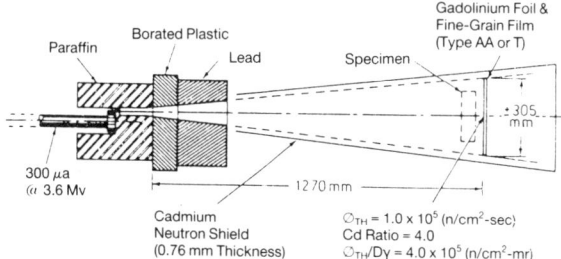

Fig 1. *Accelerator-based thermal neutron radiography system*

The last paper in the ASTM collection /6/ in which some information on collimation can be found is the paper by Kok /9/. It describes NR facilities at the BWR 2 MW swimming pool type Battelle reactor. Four neutron collimator facilities are available there. Three of these are located in the reactor pool (see fig. 2). The P-1 collimator is fabricated from a single piece of square aluminium tubing which is closed at the ends by welding on a 3.2 mm thick aluminium end cap.

The P-2 collimator is a divergent type, fabricated in sections, joined with bolted gasketed flanges. A shutter made of 12.5 mm boral is used to control exposure. A box with an open bottom is built around the exit aperture. Water is displaced from the box by filling it with helium.

The P-3 collimator is fabricated of aluminium. The shutter is made of cadmium and indium.

The dry facility utilizes a beam tube to extract a neutron beam from the reactor. The aperture is formed by a hole in a gadolinium foil. Bismuth is used as gamma shielding.

No information could be found in /9/ about the cladding of the collimators.

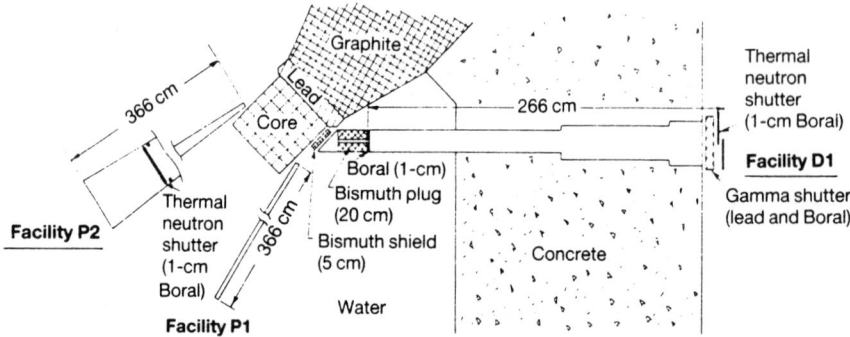

Fig. 2. Lay-out of the neutron radiographic facilities at the Battelle Research Reactor

The next publication describing many aspects of NR is the IAEA issue from 1977 of the Atomic Energy Review dealing entirely with NR problems /10/.
One could expect to find some information on collimation in the paper of Hawkesworth /11/, which according to the author concentrates on the practical side of NR. Unfortunately, this problem is not discussed in the paper.

Although the next paper, (by Ross) /12/ in the IAEA review /10/ gives only a sketch and a photograph of the underwater collimator at the pool-type High Flux Reactor at Petten, it contains short descriptions of facilities for reactor fuel neutron radiography in different countries. Details about those facilities can easily be found by looking into the ample list of references, given in /12/, referring to those facilities. This was actually done and the result of this search will be reported in 4 below.

Another paper in /10/ in which one could expect to find some information about the collimation problems is that of R.L. Tomlinson /13/. Unfortunately, although very detailed information is given on special equipment, techniques and accessories, none is given about the design and construction of the collimator of the Aerotest Radiography and Research Reactor.

Although detailed information about the design and construction of collimators for NR is not evident in the publications reviewed above, the "Neutron Radiography Handbook" /2/ prepared by the NRWG and published in 1981 has addressed the subject.
The problem of collimation is presented there by R. Matfield in two separate chapters. The first one deals with aspects of collimation itself and gives only some general information about the collimator design.

It notes that the divergent collimator is widely used for NR and specifies boron, cadmium, dysprosium, europium, gadolinium and indium as lining materials for the walls of the collimator. Other general information about the collimator design is that for low intensity neutron sources it is necessary to leave an unlined section at the beginning of the collimator, whereas in most designs of high-flux reactor sources the collimator is lined along its full length. For low-intensity sources the unlined length is usually about two diameters of the inlet aperture of the collimator.

Of additional practical interest is that long and wide collimators have better neutron to gamma ratios and hence better contrast for the direct method. The neutron attenuation due to the collimator atmosphere amounts to about 5% per meter and can be reduced to less than 1% when helium is used.

The problem of the collimator design is again treated by Matfield in a chapter devoted to the design of neutron radiography equipment in /2/. Here the problem of the selection of a material to line the walls of the collimator is discussed. The nuclear effectiveness of boron (in the form of boral), cadmium, dysprosium, europium, gadolinium or indium can be assessed for different neutron energies. Matfield concludes that "it is thus clear that there are no outstanding materials from the neutronic viewpoint and so the cost and the mechanical properties of these materials must also be considered when making a selection". The properties of interest for the lining materials mentioned above are tabulated in /2/.

The characteristics of the lining materials are also given in /2/, especially for boron and boral, cadmium and europium. The inlet aperture of the collimator must be well defined. As stated by Matfield in /2/: "As the resolution of the collimator is a function of the inlet-aperture size it is important that this aperture should be well defined. This can be achieved by constructing the inlet face of the collimator from a material which is opaque to neutrons, and especially those neutrons to which the converter foils are most sensitive". Matfield concludes that a combination of cadmium and indium will be best for the front face of the collimator. In the "Neutron Radiography Handbook" /2/, the following data are tabulated for NR installations in the European Community: collimation ratio, inlet diaphragm dimensions, collimator lining, and beam dimensions at object plane.

The Handbook /2/ also gives sketches of NR facilities, but no additional information on collimation than that listed in the above-mentioned table.

The most recent source of information reviewed for details of collimation problems in NR is the second edition of the "Nondestructive testing handbook",/14/, published in 1985 by the American Society for Nondestructive Testing. In this handbook two sections are devoted to NR.

In section 12, written by Berger et al. /15/, a special chapter is devoted to neutron collimation, considered by the authors as "an important part of neutron radiography's source technique". They thereafter confirm that "nearly all of the newer facilities use divergent collimators". Unfortunately, no specific information is given in /15/ either about the design and construction of a collimator or its materials.

Section 13 in the NDT Handbook /14/, prepared by Barton /16/ deals with implementation of NR. It does not contain any information on collimator design and construction.
Some time ago Kodak-Pathé published two booklets devoted to neutron radiography in the series on "Radiographie et industrie". They are both called "La neutronographie" /17/, /18/. Only in /18/ is the date of issue given (1974).
In the first booklet /17/, containing 11 papers, principles, installations, recorders

and general problems of NR are treated. The first paper /19/ has a chapter devoted to collimators. It gives only the principle of a divergent collimator (see fig. 3), stating that this type of collimator is most widely used. As can be seen

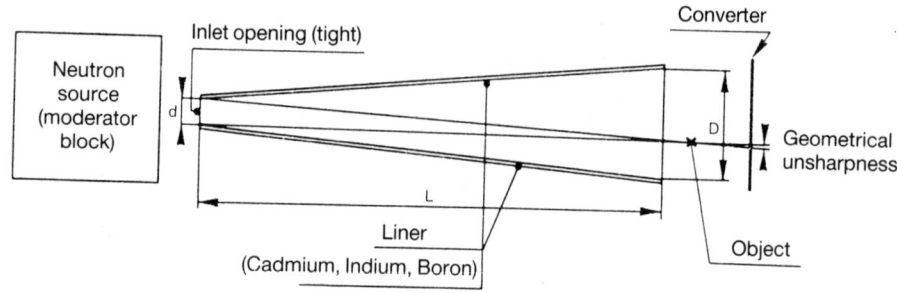

Fig.3. Divergent conical collimator

from fig. 3 the collimator is lined with cadmium, indium and boron.

The next paper /20/ in /17/ gives the principles of in-pool neutron radiography (see fig. 4). It shows also that the collimator is lined with Cd+In and B. The

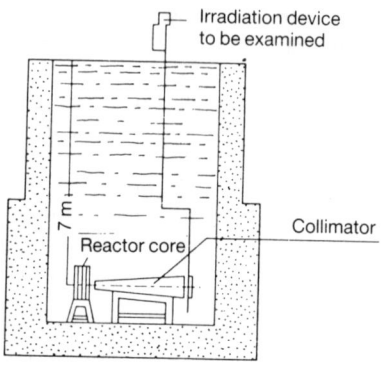

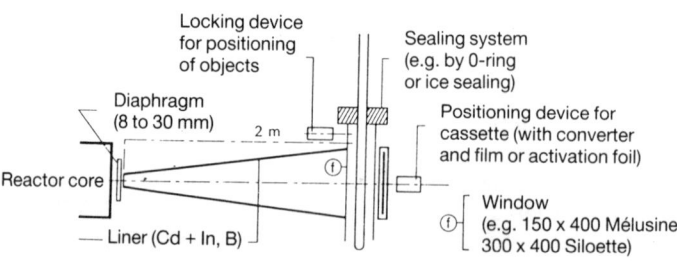

Fig. 4. Principle of in-pool radiography

dimensions of the diaphragm shown in fig. 4 are taken from the Siloette reactor (Grenoble).
In fig. 5 the NR facility at the Triton reactor (Fontenay-aux-Roses) is shown. Here

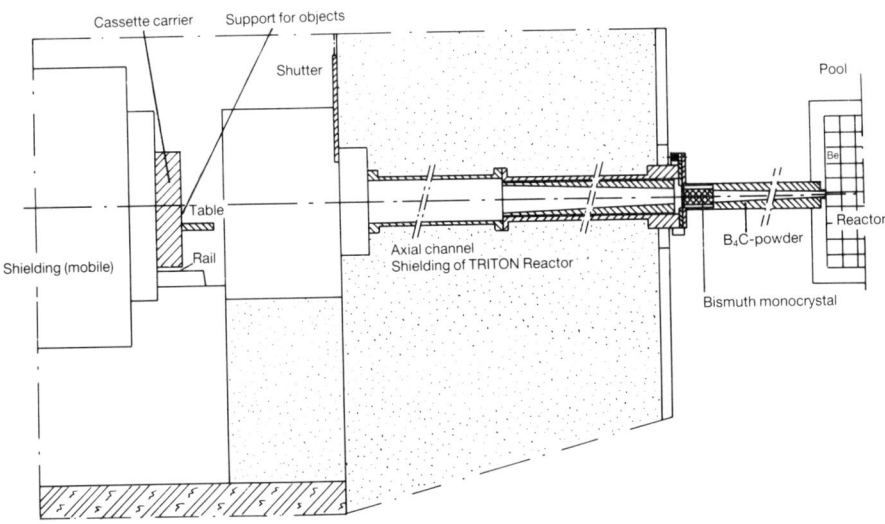

Fig. 5.   NR facility at the Triton reactor

the gamma-ray filter is made of a bismuth monocrystal (15 cm thick at the Triton and 10 cm thick at the Siloette reactor). The neutron guide between the filter and the exposure zone, has a lining of boron or lithium and is filled with helium.

Further description of the Triton facility is given in /21/ where it is stated that the collimator, filled with helium, is lined with B_4C.
In the next paper /22/ in /17/ a description is given of a mini-reactor for NR (Valduc). Here both the axial and tangential divergent collimators (see fig. 6) are lined with boral.

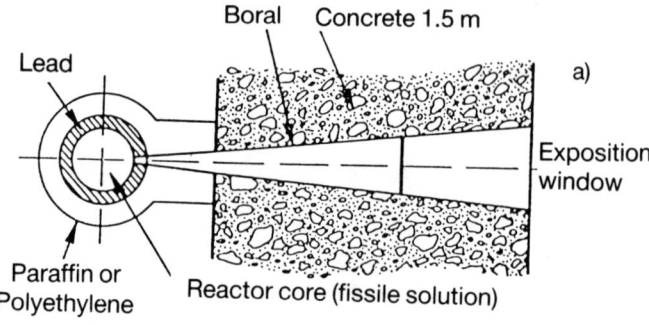

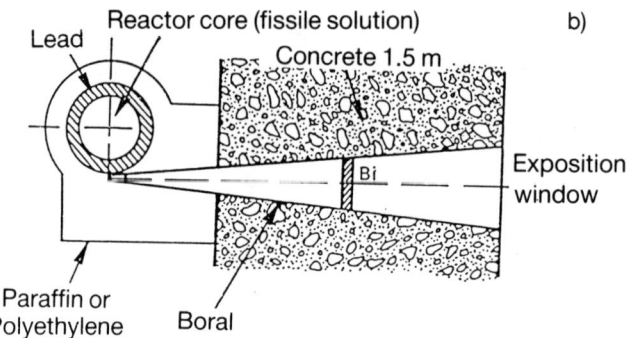

Fig. 6. Collimators of the mini-reactor. a-axial, b-tangential

The second Kodak booklet /18/ on NR deals with applications of neutron radiography and does not cover collimator design and construction.

In the search for information on collimator design and construction in handbooks and general publications about NR only reference /2/ gives some particular information. The ample reference list in /12/ can serve as a guide to look for such information in detailed publications about NR.

It was concluded that a detailed literature search was necessary in order to establish a sound knowledge base which would prove useful to those setting up or operating NR facilities. The results thereof are reported below.

3. CONFERENCES ON NDT AND NR

As it was not possible to find enough detailed and specific information about the design and construction of collimators for thermal neutron radiography in general publications on NR it was necessary to search the literature in the numerous papers presented at international conferences on nondestructive testing and neutron radiography.

At the *4th International Conference on Non-Destructive Testing* (London, 9-13.9.1983) Berger presented a paper on NR /23/.

No mention is made in this paper about collimation problems.

During the *5th International Conference on Non-Destructive Testing* /24/ (Montreal, 1967) three papers on NR were presented at a session devoted to surface methods and neutron radiography.

The first paper /25/ analyses some problems connected with multi-slit collimating systems. In the discussion which followed the presentation of the paper, J.P. Barton points out the advantages of using divergent collimators (see also /1/).

In the second paper /26/ a description is given of the Battelle Research Reactor-BRR(2MW, pool) in which a collimated beam of neutrons is obtained by immersing a 38 mm square, air-filled aluminium tube horizontally in the reactor pool (see fig. 7).

The tube is 3.6 m long. Neutron radiography of long items is done by scanning.

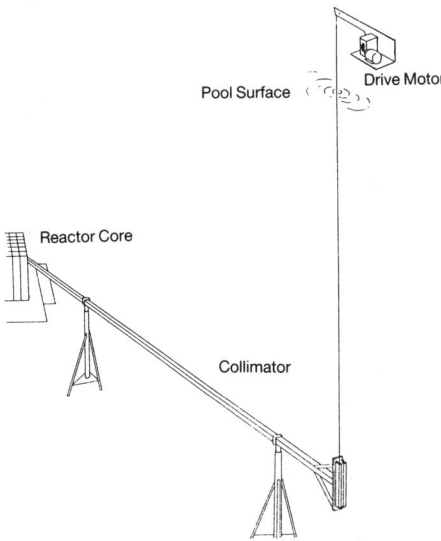

Fig. 7. NR apparatus at the BRR

The third paper /27/ presented to /24/ describes the NR facilities of the 100 kW Toshiba Training Reactor. There the collimated thermal neutron beams are obtained from two beam ports of the reactor which were simply converted to five collimators, as shown on fig. 8. One of them is constructed by pulling out the

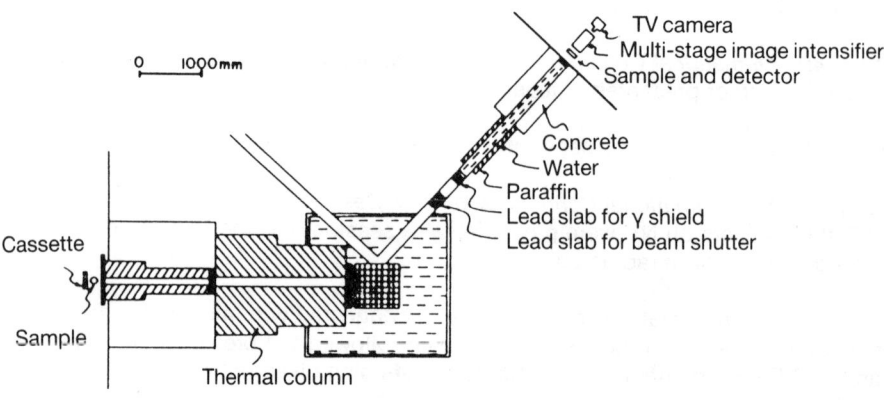

Fig. 8. Experimental arrangement for NR Toshiba Training Reactor

10×10 cm graphite rod in the thermal column and placing the collimator plug in front of it. The collimator is made by pouring parrafin into the stainless steel case. No cadmium is used, but a 50 mm thick lead block is placed at the end of the thermal column as a gamma-ray shield.

At the *6th International Conference on Non-Destructive Testing* (Hannover, 1-5.6.1970) a special session was devoted to NR and holography /28/. During this session 9 papers were presented on NR, but only in two of them were problems of collimation of thermal reactor neutron beams discussed.

In the first paper /29/, a collimating system (see fig. 9) used in the Triton (Fontenay-aux-Roses) and Melusine (Grenoble) reactors is described. The collimator (1) is situated between the core of the reactor (2) and its concrete shield (3).

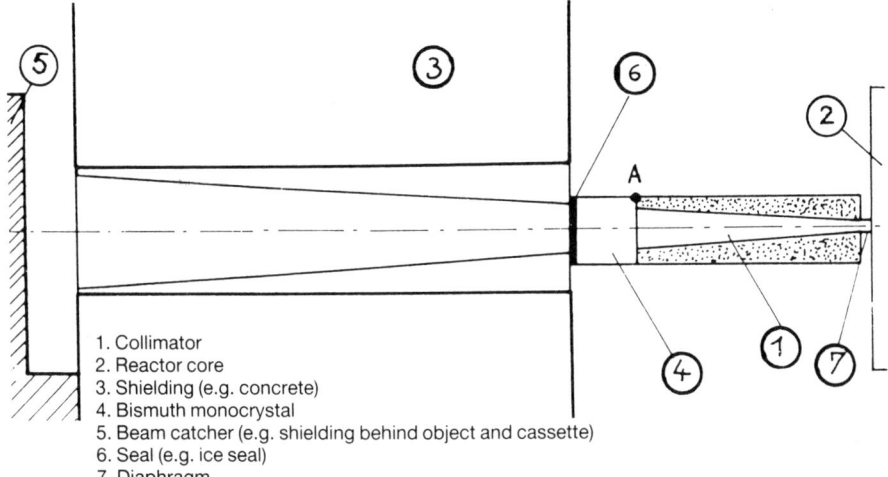

1. Collimator
2. Reactor core
3. Shielding (e.g. concrete)
4. Bismuth monocrystal
5. Beam catcher (e.g. shielding behind object and cassette)
6. Seal (e.g. ice seal)
7. Diaphragm

Fig. 9. Collimating system /29/

The B_4C powder, sandwiched between two aluminium tubes, conical at the centre and cylindrical at the outside, assures the collimation. A steel sleeve diminishes gamma background, which could reach the bismuth monocrystal (4) from outside the collimator. An ice seal (6) prevents water from entering the system. A special device (7) at the entrance of the collimator prevents a decrease in neutron flux.

According to /30/, in all French facilities of the in-pool type a collimating system as shown in fig. 4 is used. The collimator consists of a water-tight aluminium body, lined with boron or cadmium and indium. Removable diaphragms are placed at the entrance to the collimator. The collimator has a length of at least 2 m.

Three types of watertight joints are used (see fig. 10):

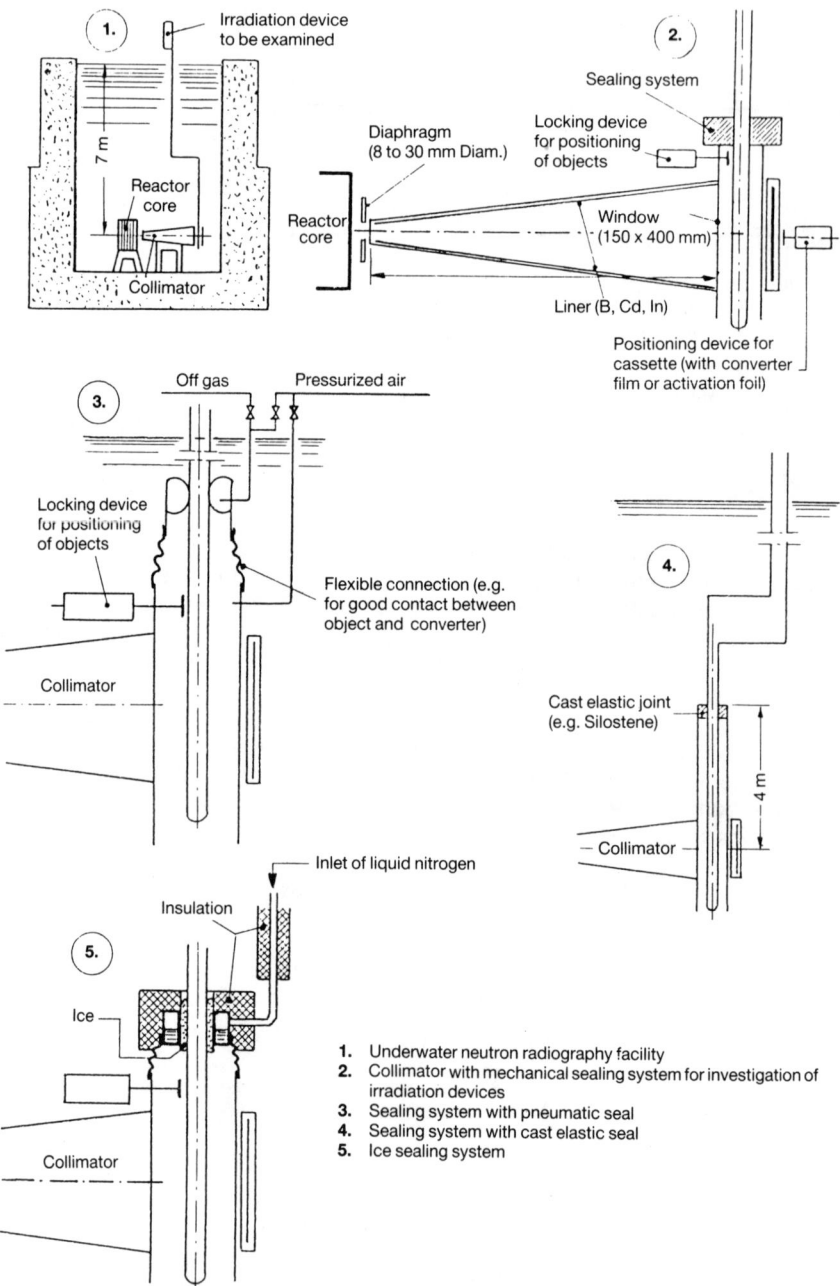

Fig. 10. *Watertight joints elastic, cast with silostene and an ice joint.*

The collimating system used for (cold) subthermal neutron radiography is shown in fig. 11.

A special session was devoted to neutron radiography at the *7th International Conference on Non-Destructive Testing* (Warsaw, 4-8.6.1973) at which 4 papers were presented. No details were given about collimators.

In /31/ it is noted that in the Hungarian reactor divergent collimators are used for NR.

Although at the *8th World Conference on Non-Destructive Testing* (Cannes, 6-11.9.1976) 10 papers were presented devoted to NR none of them contained any information about collimation problems.

At the following meeting, the *9th World Conference on Non-Destructive Testing* (Melbourne, 1979), altogether 5 papers were presented dealing with NR. Only in one of them /32/ was a description given of the collimating systems used in the axial and lateral beams of the Triton (Fontenay-aux-Roses) reactor (to be described later).

At the *10th World Conference on Non-Destructive Testing* (Moscow, 22-28.8.82) there was no special session devoted to NR. Nevertheless, 5 papers dealt with the problem. In none of them were the problems of collimation discussed.

During the *11th World Conference on Nondestructive Testing,* (Las Vegas, USA, 1985) only two papers were presented on neutron radiography. They did not describe collimator details.

Other sources of information on NR are the proceedings of the European Conferences on NDT.

General arrangement of NR facility for industrial applications at MELUSINE reactor

NR facility (with Bismuth monocrystal of 100 mm Ø x 150 mm) providing thermal neutrons

Removable filters for subthermal NR

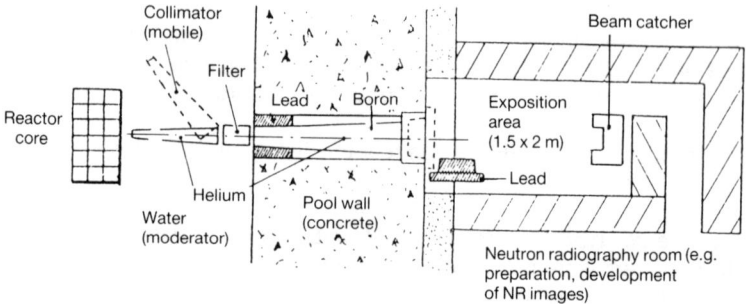

Fig. 11. Collimating system for subthermal (cold) neutron radiography.

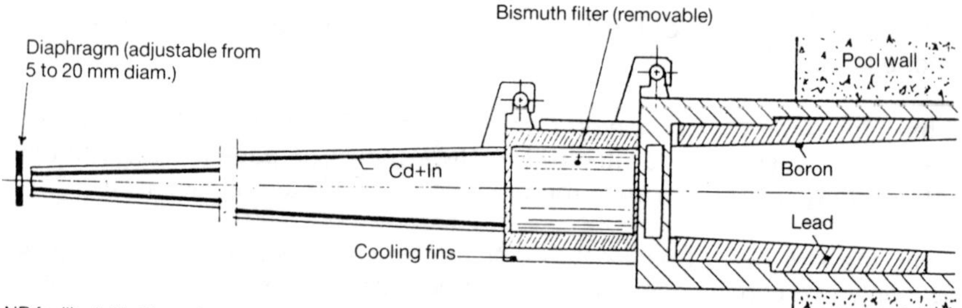

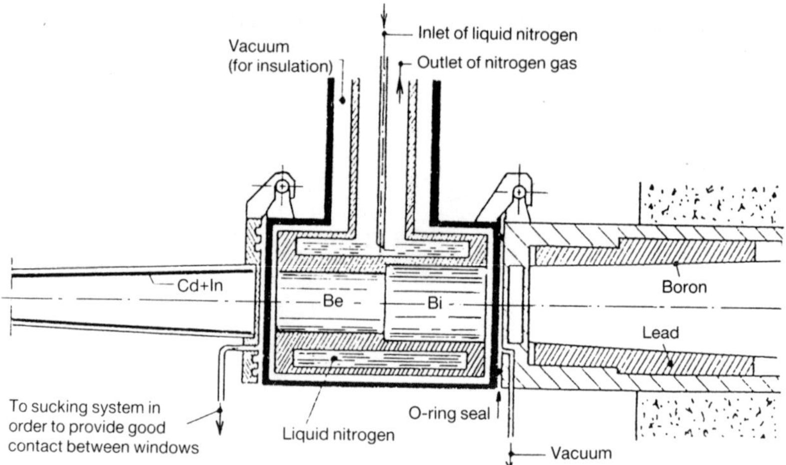

During the *1st European Conference on Non-Destructive Testing* (Mainz, 24-26.4.1978) A. Laporte described the application of the Triton reactor to NR /33/. Here again the collimation system (see fig. 5) was described (details later).

At the *2nd European Conference on Non-Destructive Testing* (Vienna, 14-16.9.1981) A. Laporte described the new NR facility at the Orphée reactor at Saclay /34/, giving, however, no details about the collimating system.

During the *3rd European Conference on Non-Destructive Testing* (Florence, 15-18.10.1984) two papers were related to NR, but did not discuss collimation problems.

Looking into other sources of information about NR, as an NDT method, papers presented at International Conferences on Non-Destructive Evaluation in Nuclear Industry must be considered. One can indeed find there some papers dealing with NR.

At the *4th International Conference on Non-Destructive Evaluation in Nuclear Industry* (Lindau, 25-27.5.1981) one paper on NR was presented.

At the *6th International Conference on Non-Destructive Evaluation in Nuclear Industry* (Zürich, 28.11-2.12.1984) five papers describe the use of NR.

At the *7th International Conference on Non-Destructive Evaluation in the Nuclear Industry* (Grenoble, 28.1-1.2.1985) two papers dealt with NR.

At the *8th International Conference on NDE in Nuclear Industry* two papers were devoted to NR, but did not discuss colimation problems.

Unfortunately, in the proceedings from these conferences no indications about collimation problems could be found. All the papers dealing with NR concentrated on the applications of the NDT method and not on the details of the facilities.

At some other international symposia and conferences the subject of NR was discussed. As it is impossible to review all such events, only a few examples are given below.

At the IAEA *Symposium on Irradiation Facilities for Research Reactors* (Teheran, 6-10.11.1972) C. Desandre-Navarre described NR performed with intermediate power reactors /35/.

In his review paper Desandre-Navarre gives some general indications about collimators used in immersed and "dry" facilities.

In the immersed facilities divergent, conical collimators are used. The walls of the collimator are made of aluminium and are lined with boral (as B_4C) or with cadmium and indium. The use of boron is recommended because it becomes less activated, which facilitates maintenance of the collimator.

The "dry"-type facilities are particularly suitable for industrial applications. As an example, in /34/ a description is given of the Triton reactor (shown previously in fig. 5). Similar installations are used in the Melusine and Siloette reactors. The

Triton NR facility is described in several publications, which will be reviewed later.

In /34/ general recommendations are given for the design of a collimator used in a 1 MW swimming-pool reactor. A part of a tangential channel is located in the water of the pool and the rest in the concrete shield of the reactor. The collimator consists of a B_4C cylinder inside an aluminium container. At the entrance a circular diaphragm controls the neutron flux and the sharpness of the neutron image. The B_4C powder, placed between two aluminium cylinders, has a thickness of between 15 and 20 mm, depending on the powder granulation. It is recommended to have at least 1.85×10^{22} boron atoms per cm².

The B_4C cylinder extends up to the second intermediary disc (also from B_4C), located at the bismuth crystal. This crystal, placed at the pool side, is encased in aluminium in contact with the wall of the channel to assure good cooling conditions.

A divergent protection screen is located at the rear of the bismuth crystal to stop the gamma radiation atenuated by the bismuth.
At the exit of the channel a port of borated paraffin and lead is used to close the channel during maintenance operations.

In 1976 the IAEA organized a *Seminar on Nuclear Quality Assurance* (Oslo, 24-27.5.1976) /36/. At this seminar the Belgian Mol BR1 reactor, used for NR, was described /37/ (see fig. 12).

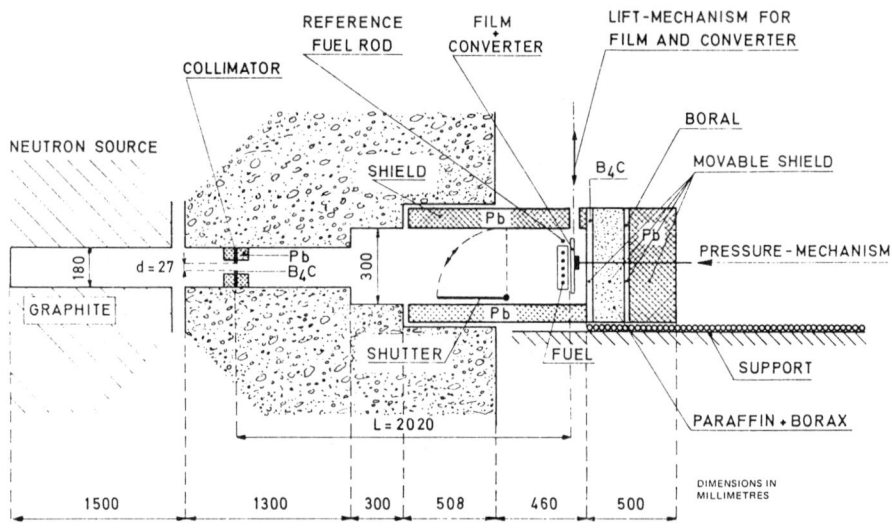

Fig. 12. Lay-out of the NR facility at BR1

The collimator is made of a 2.5 cm thick B_4C sheet with a circular aperture of 30 mm. It is mounted between two 5 cm thick lead slabs in order to reduce the gamma components in the neutron beam. To improve the neutron beam collimation, part of the channel is covered with boral. The L/D ratio was fixed at 75 as a result of a compromise between irradiation time and unsharpness.

In another paper presented at /36/ a neutron radiography facility is shown /38/, but without giving details of the collimating system (see fig. 13).

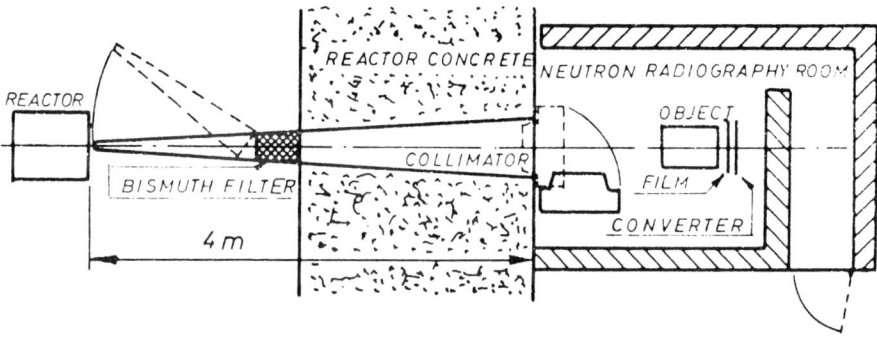

Fig. 13. Neutron radiography bench

During another International Syposium "New Methods of Non-Destructive Testing of Materials and their Application Especially in Nuclear Engineering" (Saarbrücken, 17-19.9.1979) /39/ two papers were devoted to NR. In the first /40/ the FRG-2 facility at Geesthacht was described. Details about this facility will be given later, when describing NR facilities in the FR Germany.

The second paper /41/ describes neutron laminagraphy. No details about collimator problems are given.

Information about neutron radiography can be found also in the proceedings of the *Post-irradiation Examination Conference* (Grange-over-Sands, 13-16.5.1980). /42/. The R2 neutron radiography facility at Studsvik is described in /43/ and shown in fig. 14. More details about the Studsvik facility are given later in this report.

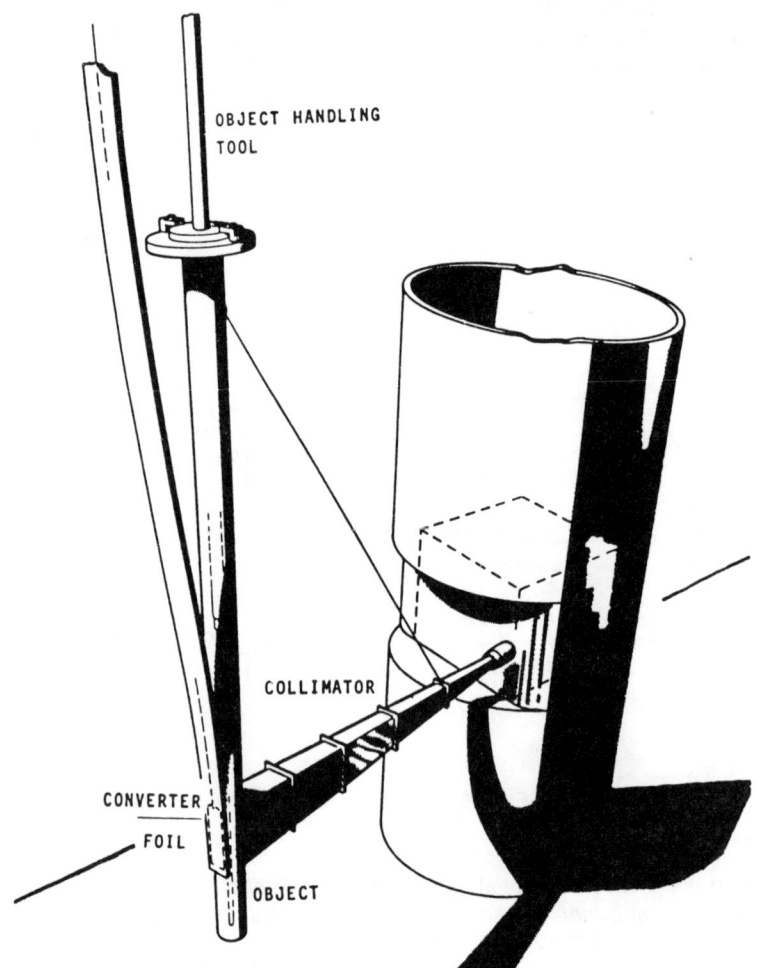

Fig. 14. NR facility at Studsvik R2 reactor

The proceedings of the *First World Conference on Neutron Radiography* (San Diego, Dec. 1981) /44/ present an invaluable source of information. At this conference 12 papers described reactor facilities with general applications and 12 papers were published on reactor facilities with nuclear applications. In several papers of the conference devoted to other topics, information can also be found about collimation problems. Altogether 19 papers from this conference are reviewed below.

In some cases the papers presented at the conference /44/ contained information published previously on the same subject or on the same NR facility. Such papers are reviewed simultaneously.

During the *2nd World Conference on Neutron Radiography* (Paris, June 1986) in many oral and poster presentations NR facilities were described. As the proceedings of the conference are not yet published, only preliminary information can be given, derived from the abstracts. Nine papers presented at session II "Reactor facilities" and 7 papers and 3 posters in session III "Non reactor sources" are a likely source of information. All papers will be published in the proceedings of the conference.

The *International Neutron Radiography Newsletter* /45/ published in English in the British Journal of Non-Destructive Testing and in French in Revue Pratique de Control Industriel publishes short descriptions of NR facilities and the activities of the NRWG members in the European Community. The INRNL is a continuation of a previously published Neutron Radiography Newsletter, edited by J.P. Barton. Fourteen issues of the NRN were published by the American Society for Non-Destructive Testing. An edited and indexed compilation of the combined back editions of the Netron Radiography Newsletter was published in December 1977 as a special issue: "Neutron radiography 1964-1977" /46/. It contains a list of more than 700 reports on NR, in which 15 references on collimation are quoted.

Up untill July, 1986, 12 issues of the INRNL /45/ were published. They give short descriptions of various NR facilities (Mol, Belgium in No 5; Petten, The Netherlands in No 6; Grenoble, France in No 7; Harwell, England in No 8; Risø, Denmark in No 9; Haifa, Israel in No 10; Ljubljana, Yugoslavia in No 11; Saclay-Orphée, France in No 12). A detailed description of collimators used in these NR facilities are given in 4 below.

4. DESCRIPTION OF NR FACILITIES

Detailed information is given below on collimation systems described in different papers, especially in /44/. This information is arranged according to the countries in which NR facilities are located.

The review given below is far from complete. It is based on information already given in chapters 2 and 3 above, on papers presented at /44/. and several other reports and papers which could be found elsewhere.

4.1. Australia

Although some information about the NR facility at Lucas Hights can be found in /47, 48/, there is no detailed description of the collimating system in use.

4.2. Belgium

The Nuclear Research Centre of Mol (SCK/CEN-Mol) has two NR facilities /49/. The dry facility of the BR-1 reactor is used for NDT control of non-irradiated fuel pins and/or materials, with the direct method. The second facility installed on the BR-2 reactor is of the pool type and is mostly used for NDT with the transfer method.

The description of the *BR-1* facility can be found in two papers from 1976. The first /37/ presented at an IAEA seminar in Oslo /36/ gives the same description as produced in a Mol report /49/. The lay-out of the BR-1 facility, shown in fig. 12, can be found in /37, 49, 51/ as well as in /2/. It is already described in 3 above.

The NR facility at the *BR-2* is described in /36, 49, 52/. The general and detailed lay-out is shown in fig. 15 /2/.

The collimation system in the form of the truncated pyramid with a rectangular base, covered with boral (B_4C) has a diaphragm diameter of 13 mm, an L/D ratio of 200, and is filled with helium.

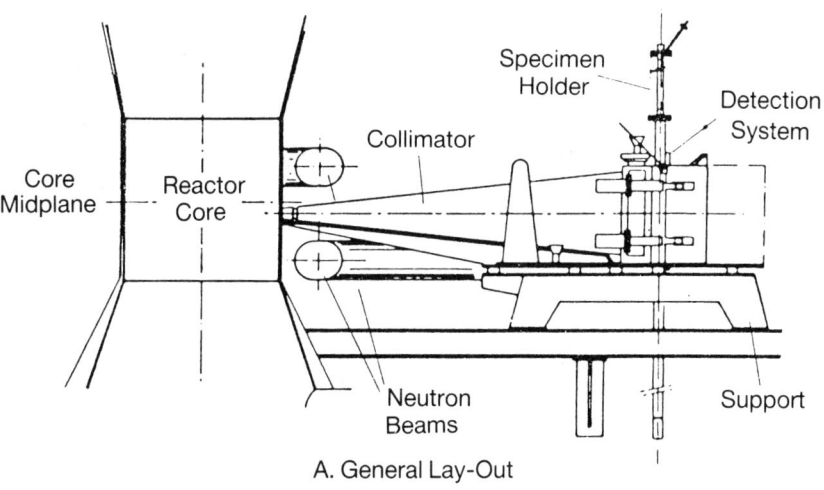

A. General Lay-Out

Facility	L_b	L_o
1	2603	28
2	2577,5	53,5
3	2595 – 2655	0 – 60
4	2605	81
5	2615	16

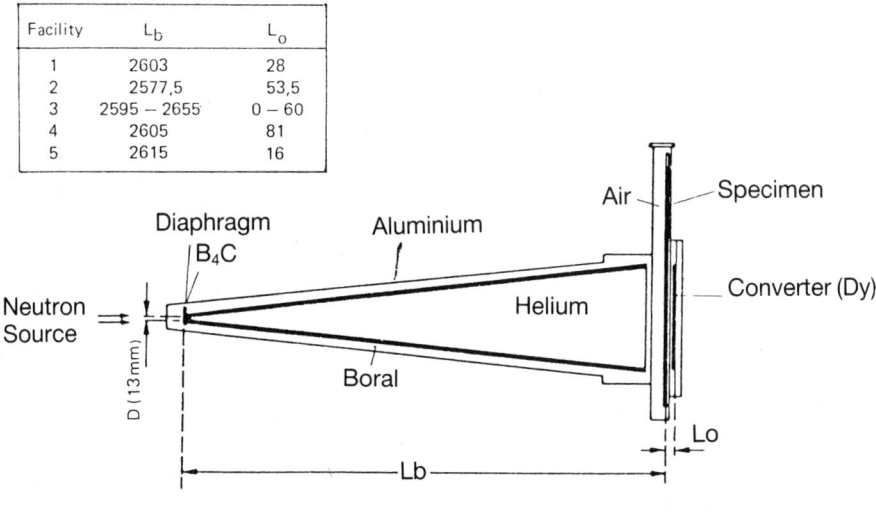

B. Detailed Lay-Out

Fig. 15. General and detailed lay-out of the BR 2 facility

4.3. Brazil

At the Centre for the Development of Nuclear Technology (CDTN) in Belo Horizonte a study was made as early as in 1974 /53/ to accomodate a NR installation at the Triga MKI Reactor (100 kW). As reported in /53/ in 1981 there were no plans for installing such a facility.

According to /53/ it was envisaged that a divergent collimator be assembled from several cylindrical parts with increasing diameters. The first part of the collimator (entrance side) need not be lined. At the entrance a lead shield (with conical opening) is placed to reduce the gamma rays entering the collimator. Several other parts of the collimator (see fig.16) are lined with boral and filled with helium.

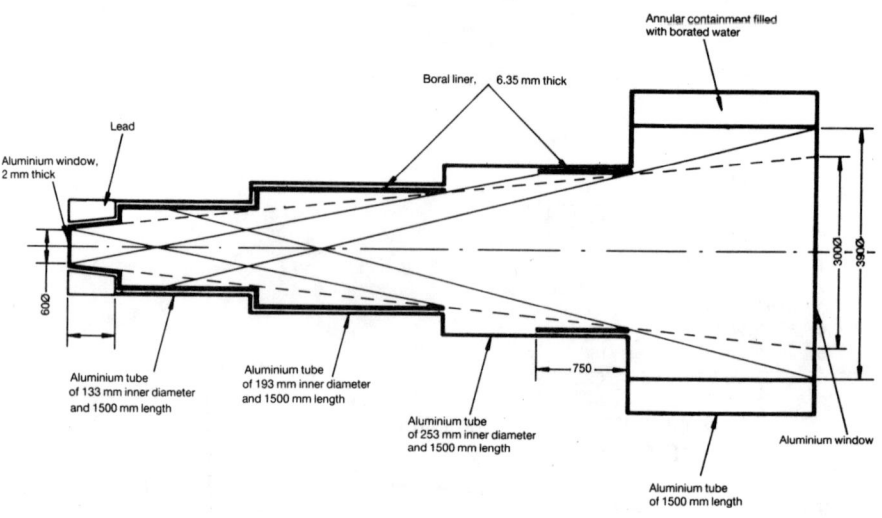

Fig. 16. Divergent collimator at CDTN

The last part of the collimator (exit side) is surrounded with borated water, which serves as a radiation screen. Apart from the above description, in /53/ interesting data can be found about the utilization of boral.

At the Institute of Energy and Nuclear Research (IPEN) in Sao Paulo NR is carried out at a 5 MW swimming pool reactor (IEA R-1) designed and built by Babcock and Wilcox Co. It was described in /54, 55, 56/. Here a divergent collimator is used (see fig.17).

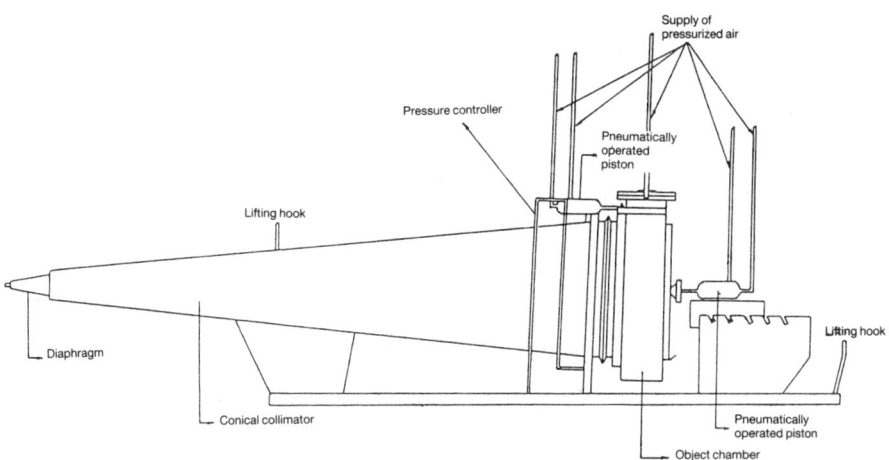

Fig. 17. Submerged NR facility at IEA-R1

A rectangular truncated pyramid is made of aluminium lined with boral. At the entrance to the collimator a small aluminium "nose" is located. It is internally lined with dysprosium, indium, gadolinium and cadmium. This "nose" was also truncated at its entrance and connected to a cylindrical aluminium tube (diameter of 8 mm), which functions as a pointlike source of neutrons. The collimator is filled with gas (of low neutron attenuation) under a pressure corresponding to that of the water surrounding the collimator.

For the Brasilian built Argonaut reactor being operated at the Institute of Nuclear Engineering in Rio de Janeiro several different types of collimators have been studied. No details of their design and construction were given in /54/.

It is mentioned in /57/ that for cold neutron NR at the Triga MkF 1.5 MW General Atomic reactor at the Federal University of Rio de Janeiro (COPPE) a stepped divergent collimator (approximately 6 m long) filled with argon, is used.

4.4. Canada

Although there were four papers from Canada presented at /44/ none of them considered collimation problems. A. Ross from AECL notes in his review paper on NR inspection of nuclear fuel /12/ that the collimator at the 30 MW NRX reactor at Chalk River is soon to be replaced. Some information about the old collimator (see fig. 18) can be obtained from /58/, where one can see that the collimator was lined with cadmium.

In the description of the NR facility used at the Mc Master University, Hamilton, Ontario /59/ the only information about the collimator is that a collimated neutron beam is obtained by placing a 6.1 m long aluminium tube vertically in the swimming pool. No other details about the collimating system are given in /59/.

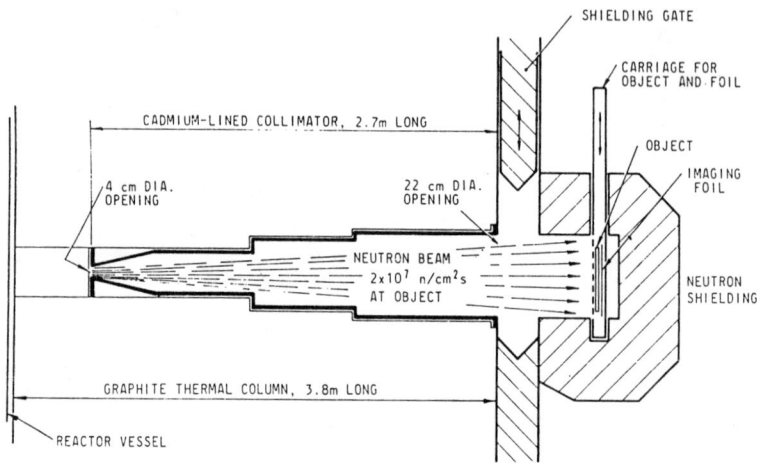

Fig. 18.   Arrangement for neutron radiography in the NRX reactor

4.5.   Czechoslovakia

The only information about the collimating system in the VVR-S reactor which can be obtained from /60/ is the shape of the collimator: it is a divergent, conical collimator with a L/D of 320.

The same type of divergent, conical collimator will be used in the SR-0 reactor /61/.

4.6.   Denmark

In the Risø double beam neutron radiography facility (see fig. 19) described in /62, 63/ the collimator is formed by removing two graphite blocks, positioned

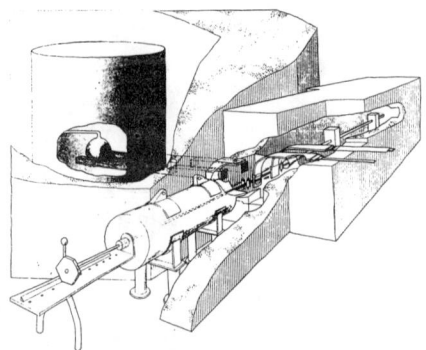

Fig. 19.   Riso double beam neutron radiography facility

tangentially to the reactor core. This was done on a length of 50 cm. At the entrance a rectangular hole of 2×8 cm is made, and the object to be radiographed is placed at a distance of 222 cm from the reactor core /64/. Thus an L/D of 110 is reached in the vertical direction and 27.75 in the horizontal direction. No internal lining is used for the collimator. At the end of the 50 cm hole a borplastic and cadmium diaphragm is used /64/.

4.7. France

There are several NR facilities operating in France. According to /2/ they are located at the following centers: Cadarache (LDAC), Fontenay-aux-Roses (Triton), Grenoble (Melusine, Siloe) and Saclay (Osiris, Isis, Orphée), Valduc (Mirene).

As early as 1972 an overview of NR facilities in France was given in /65/. A chronological list of NR facilities first constructed in 1966 appears in this report (see table 1).

Table 1. French NR facilities, constructed between 1966 and 1970 /65/.

	Operational since		Location	Reactor			Collimation
				Name	Power	Type	
1	May	1966	Cadarache	PEGGY	1 kW	Pool	Cylindrical, no shielding
2	July	1966	Grenoble	MELUSINE	4 MW	Pool	Conical, Cd + In
3	Febr.	1967	Saclay	ISIS	800 kW	Pool	Conical, B
4	May	1967	Cadarache	PEGGY	1 kW	Pool	Conical, B
5	June	1967	Grenoble	SILOE	30 MW	Pool	Conical, Cd + In
6	July	1968	Saclay	OSIRIS	70 MW	Pool	Conical, B
7	July	1968	Fontenay	TRITON	6 MW	Pool	Conical, B
8	Jan.	1969	Fontenay	TRITON	6 MW	Pool	Conical, B
9	March	1969	Cadarache	PEGASE	30 MW	Pool	Conical
10	May	1969	Grenoble	MELUSINE	4 MW	Pool	Conical, B
11	May	1969	Grenoble	MELUSINE	4 MW	Pool	Conical, B
12	Febr.	1970	Grenoble	MELUSINE	4 MW	Pool	Conical, Cd + In

In all the above facilities the underwater NR is based on the same principle, as shown on fig. 20 /2, 65/.

At the NR facilities not only underwater NR is performed, but in some instances the neutron beam is guided out of the reactor by beam tubes for "dry"-NR.

As mentioned before, in /2/ a list of NR facilities in France, existing in 1981, is given. Since then the Triton reactor is no longer used and is replaced by the NR facility at the Orphée reactor.

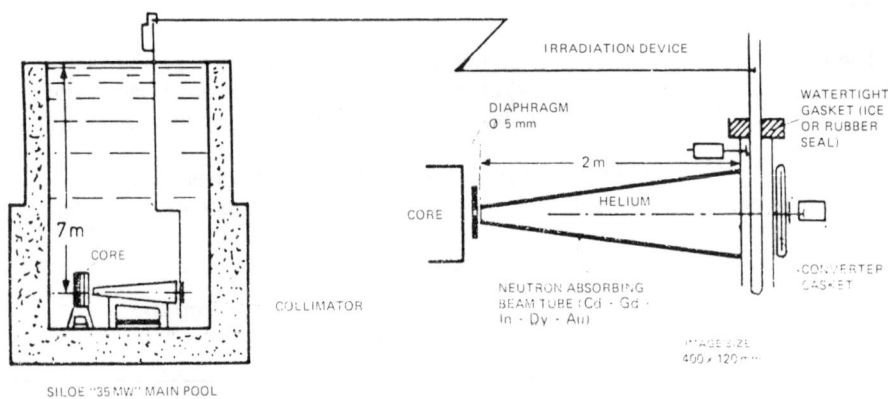

Fig. 20. Underwater neutron radiography

The collimating systems used in the French NR facilities will be reviewed in detail below.

The NR facility at the *Triton* reactor at Fontenay-aux-Roses (no longer in operation) is described many times in different papers and reports.

The NR facility at the axial channel of the Triton reactor is shown in this report in fig. 5, which is reproduced from /20/ in the Kodak-Pathé booklet on NR /17/. The same collimator is also described in /34/. The same drawing is also given in /32, 33, 66, 67, 68, 69, 70/ and is also reproduced in /2/.

Neutron radiography is also done on the lateral channel of the Triton reactor. (See fig. 21 taken from /2/). This facility is described in /32, 33, 68, 69, 70/. The collimator used there is shown in fig. 22. It is of a biconical type, with variable L/D ratio (110 to 760).

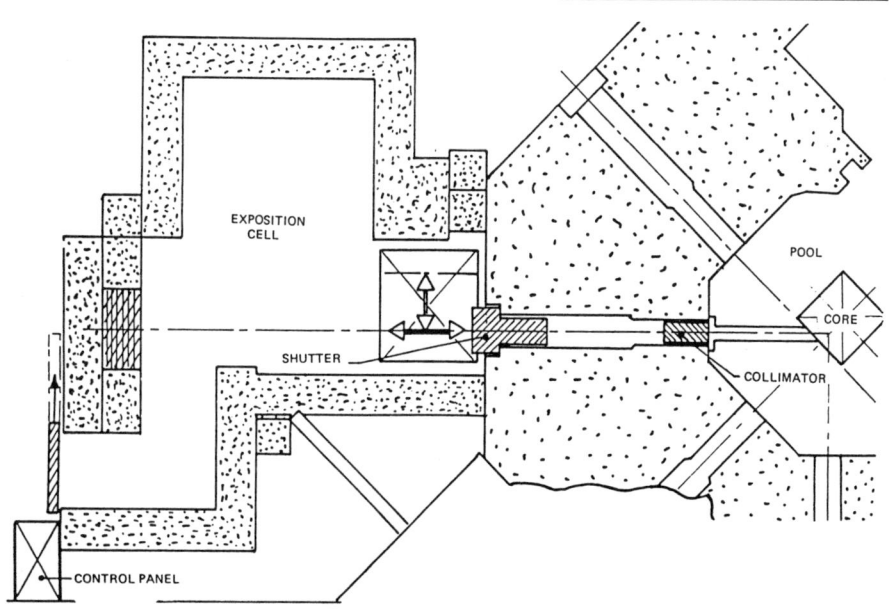

Fig. 21. NR facility at the lateral channel of the Triton reactor

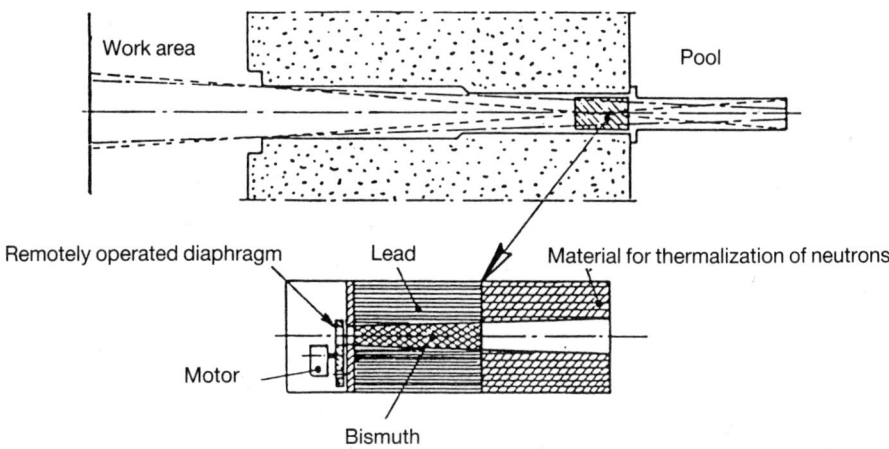

Fig. 22. Collimator of the laterar channel NR facility of the Triton reactor

The in-pool NR facility of the *Osiris* reactor at Saclay is shown in fig. 23 /71/. It consists of an aluminium pyramid lined with B_4C.

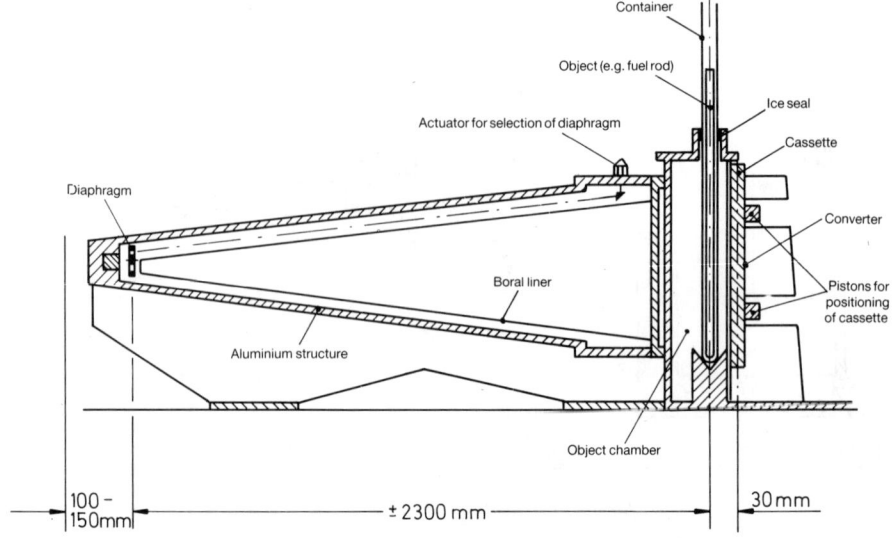

Fig. 23. Collimator of the Osiris facility

An interesting description of the evolution of the collimating system in the *Isis* reactor is given in /69/ (see fig. 24).

The first collimator (A) installed in the Isis reactor in 1966 consisted of a truncated aluminium cone lined with cadmium. Next, a truncated aluminium pyramid is used (B) lined with B_4C, having a B_4C-Al diaphragm.

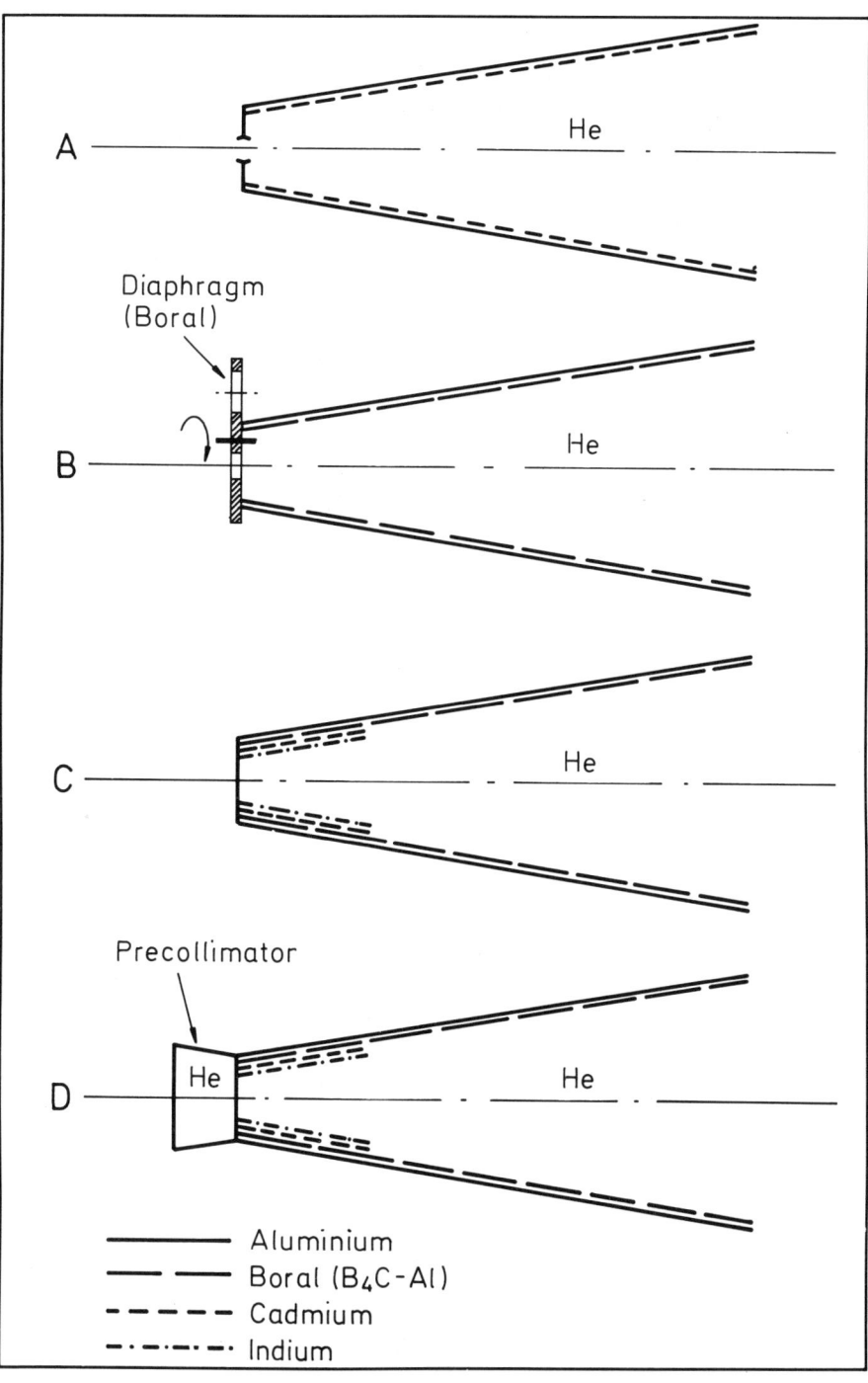

Fig. 24. Evolution of the Isis collimating system

To improve the quality of the neutron radiographs the variable diaphragm is removed and the first part of the collimator is lined with B_4C-Cd-In (C).

Finally a pre-collimator is added (D) thus arriving at a convergent-divergent geometry. This is shown in more detail in fig. 25 /2/.

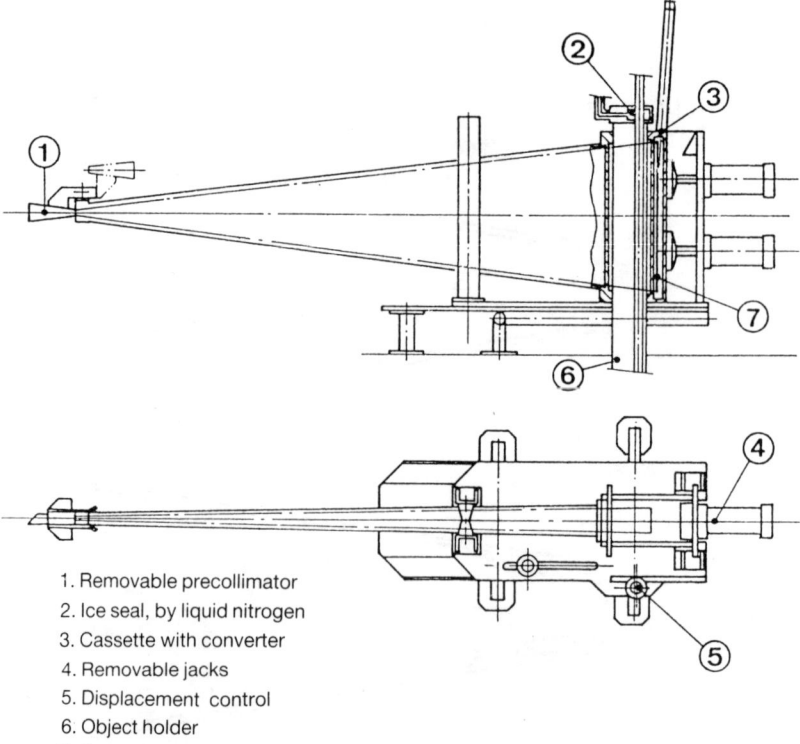

1. Removable precollimator
2. Ice seal, by liquid nitrogen
3. Cassette with converter
4. Removable jacks
5. Displacement control
6. Object holder
7. Converter

Fig. 25. Collimators of the Isis and Osiris facilities at Saclay.

For post-irradiation neutron radiography of long nuclear reactor fuel rods the Isis NR facility is developed at Saclay. It is shown in fig. 26 /2/. It is described in /66, 69, 72/.

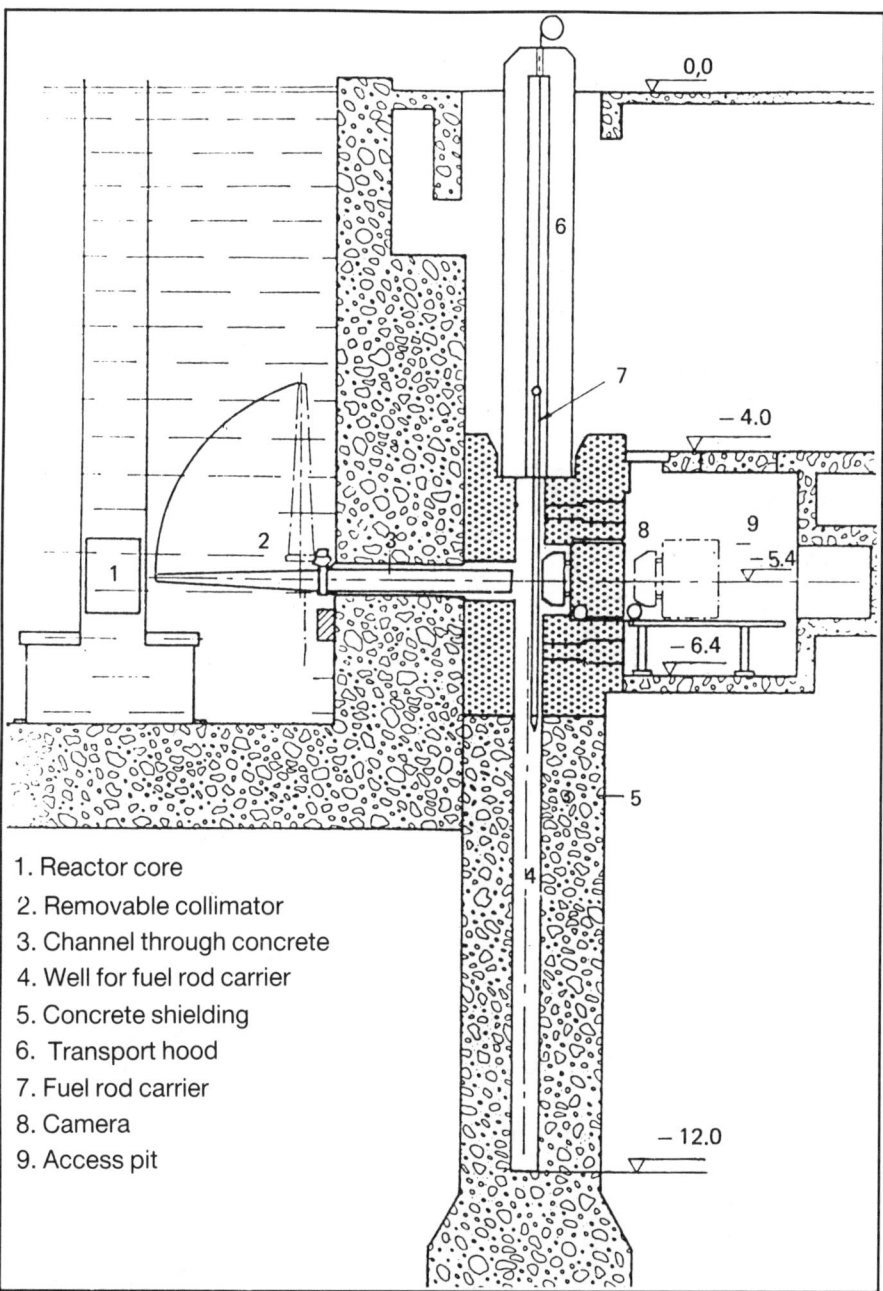

Fig. 26. The "dry" NR facility at Isis reactor at Saclay

1. Reactor core
2. Removable collimator
3. Channel through concrete
4. Well for fuel rod carrier
5. Concrete shielding
6. Transport hood
7. Fuel rod carrier
8. Camera
9. Access pit

All the French NR facilities reviewed above are also described in /73/.

The newest NR facility at Saclay is located at the *Orphée* reactor (swimming-pool type, 14 MW) (See fig. 27). The exposure room is located at about 70 m from the reactor. According to /69/ in this NR facility there is no classical divergent or converget-divergent collimator. The collimation is assured by the outlet of the neutron guide with the following parameters: dimensions of the exit beam — 25×150 mm; expected neutron flux — 5.10^8 n.cm^{-2}.s^{-1}; angle of the diverging beam — 12′ to 14′. No further details about the neutron guide are given.

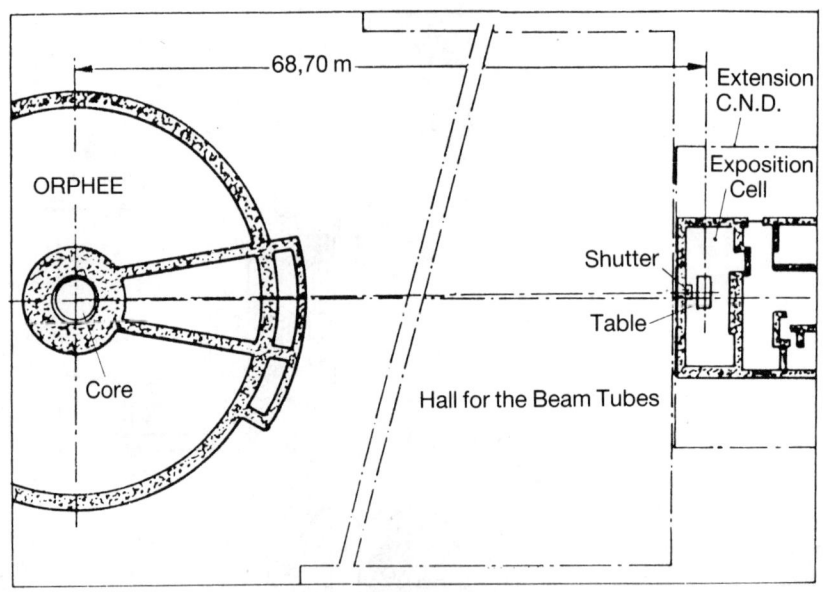

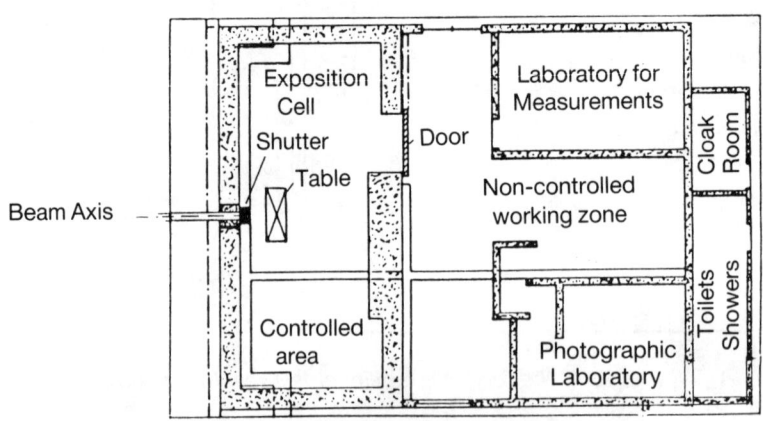

Fig. 27. NR facility at the Orphée reactor at Saclay

The Orphée NR facility is further described in /74/. In the exposition cell (see fig. 27) a post-collimation unit in the form of a 1.5 m long vacuum tube, placed at the shutter exit, is used. At the end opposite the shutter, it is fitted with a gadolinium diaphragm system used to alter the effective beam dimensions.

Before J.P. Barton had advertised the use of divergent beam collimator for neutron radiography /1/ he has tested this design in the 4 MW pool reactor *Melusine* at Grenoble /75/. The form selected for the collimator (see fig. 28) was that of a single hollow tube to be air filled and submerged to a horizontal

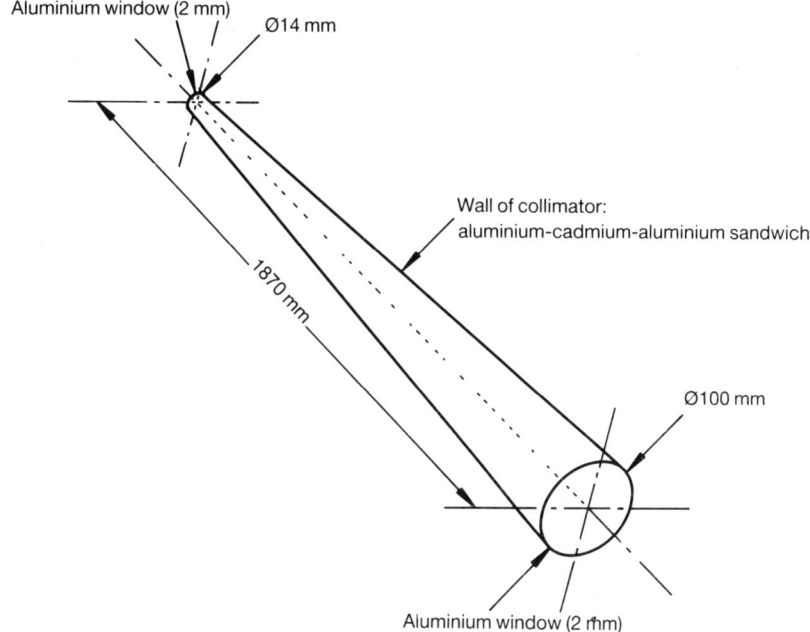

Fig. 28. Conical collimator

position pointing radially at the reactor core. The collimator was constructed as three cones placed one inside the other; the outer and inner cones being of aluminium 2 mm thick and the middle sandwiched cone of cadmium 1.5 mm thick. The windows at either end of the collimator were chosen to be of aluminium also 2 mm thick, this being sufficient to withstand the water pressure (8 m), but producing negligible neutron absorption or scattering.

The present NR facilities at Grenoble are described in /76/. The *Siloe* (35 MW), *Melusine* (8 MW) and *Siloette* (100 kW) are compact, open-core, pool-type reactors. The principle of using those reactors for underwater radiography was already shown in figs. 4 and 20 and also described in /30/. In /77/ the in-pool NR facility of the *Siloe* reactor is described and its principle is shown (see fig. 20).

In /77/ the collimator is described as follows: Divergent collimator, with rectangular cross section, can be moved back and forth by 30 cm. The body of the collimator is made of B_4C and In and the collimator's nose of (Cd+In+Gd+Au+Dy). The collimator opening is ⌀ 6 mm.

The description of the *Melusine* "dry" NR facility is also given in /77/. Unfortunately, not much information can be extracted from fig. 29 (reproduced also in /2/).

According to /77/ the convergent-divergent collimator has B_4C+In mobile diaphragms: ⌀ 16.2 and 50 mm with a variable L/D from 125 to 400. It has a permanent gamma filter of a 150 mm bismuth monocrystal.

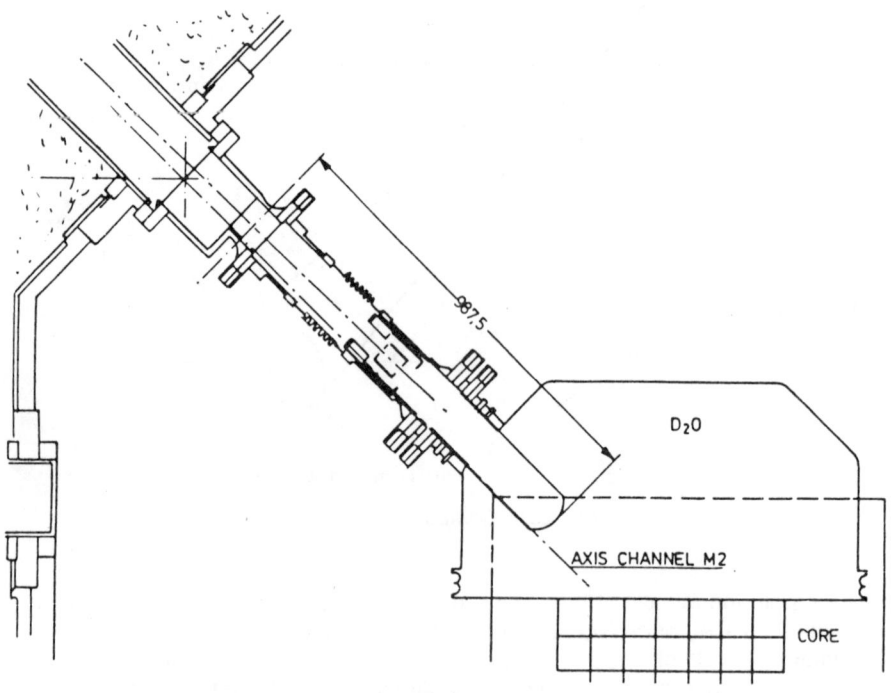

Fig. 29. NR facility at the Melusine reactor at Grenoble

It was already mentioned in /29/ that a pulsed NR facility is designed and constructed in France. It is described in /22/ and in a paper presented at the BNES conference "Radiography with neutrons" (Birmingham, 10.9.1973) /4/ and in a special report /78/. The general lay-out for this mini-reactor for NR (at Valduc) is given in. fig. 30. As can be seen radial and tangential collimators are available for NR (see also /79/). Pyramidal collimators are used, lined internally with boral.

They are 1.8 m long, have an inlet aperture of 2×3 cm and an exit aperture of 18×24 cm.

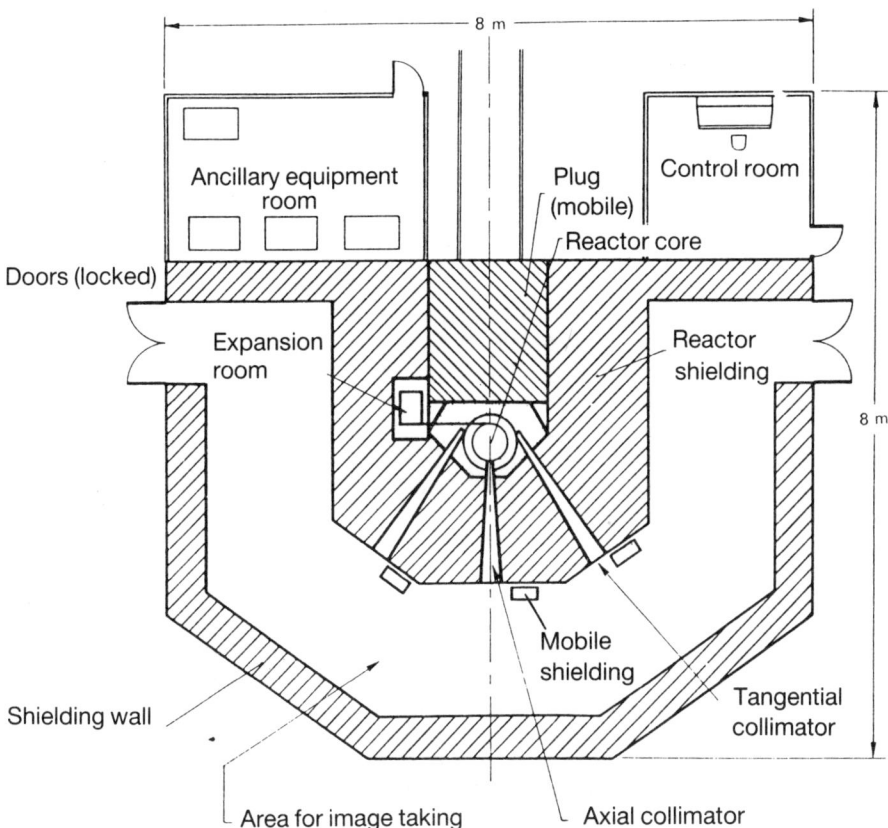

Fig. 30. Lay-out of the mini-reactor for NR

The mini-reactor *Mirene* is also described in a special pamphlet /80/, from which fig. 31 is reproduced. (reproduced also in /2/).

Similar mini-reactors for NR were also constructed at Cadarache and will be described below.

The NR facility at the *Rapsodie* reactor at Cadarache is described in /81/. At the beginning a multichannel collimator was tried. It did not meet the expectations and therefore was superseded by a divergent collimator in the shape of a truncated pyramid (entrance – 12×36; exit – 130×500 mm), lined with sheets of polyethylene (10 cm thick) and cadmium (0.8 mm thick) /78/ (where entrance and exit of the collimator is given as 3×7 and 10×50 cm).

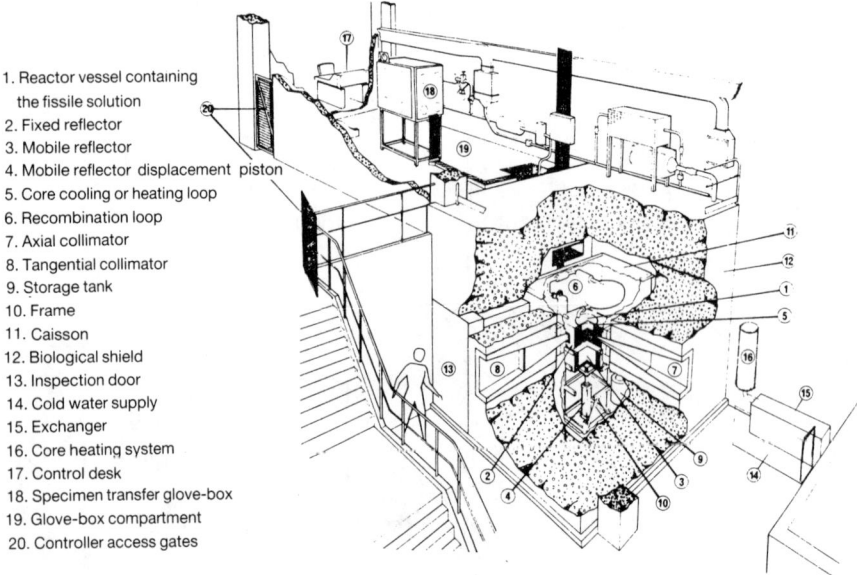

1. Reactor vessel containing the fissile solution
2. Fixed reflector
3. Mobile reflector
4. Mobile reflector displacement piston
5. Core cooling or heating loop
6. Recombination loop
7. Axial collimator
8. Tangential collimator
9. Storage tank
10. Frame
11. Caisson
12. Biological shield
13. Inspection door
14. Cold water supply
15. Exchanger
16. Core heating system
17. Control desk
18. Specimen transfer glove-box
19. Glove-box compartment
20. Controller access gates

Fig. 31. Cutaway view of Mirene

A recent description of te NR facilities at Cadarache can be found in /81/. There, the NR facilities at the LDAC (Rapsodie) and CEI (Phenix) (at Marcoule) reactors are schematically shown (see figs. 32 and 33).

Both reactors use a divergent collimator consisting of a truncated pyramid canal of polyethylene walls with 0.8 mm cadmium lining /82/.

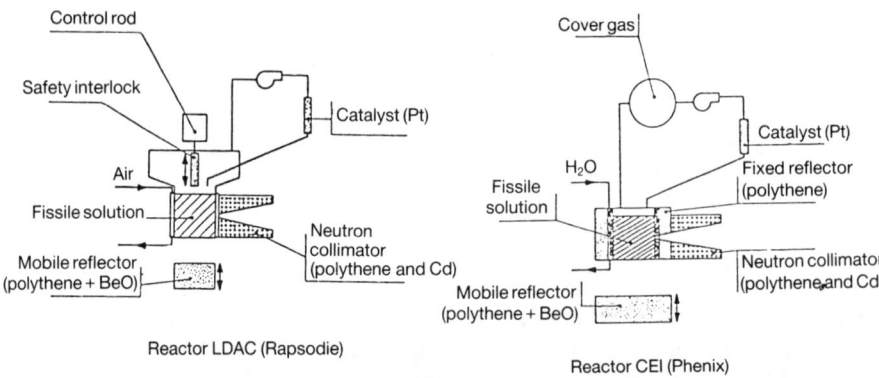

Fig. 32. NR facilities at the Rapsodie and Phenix reactors at Cadarache

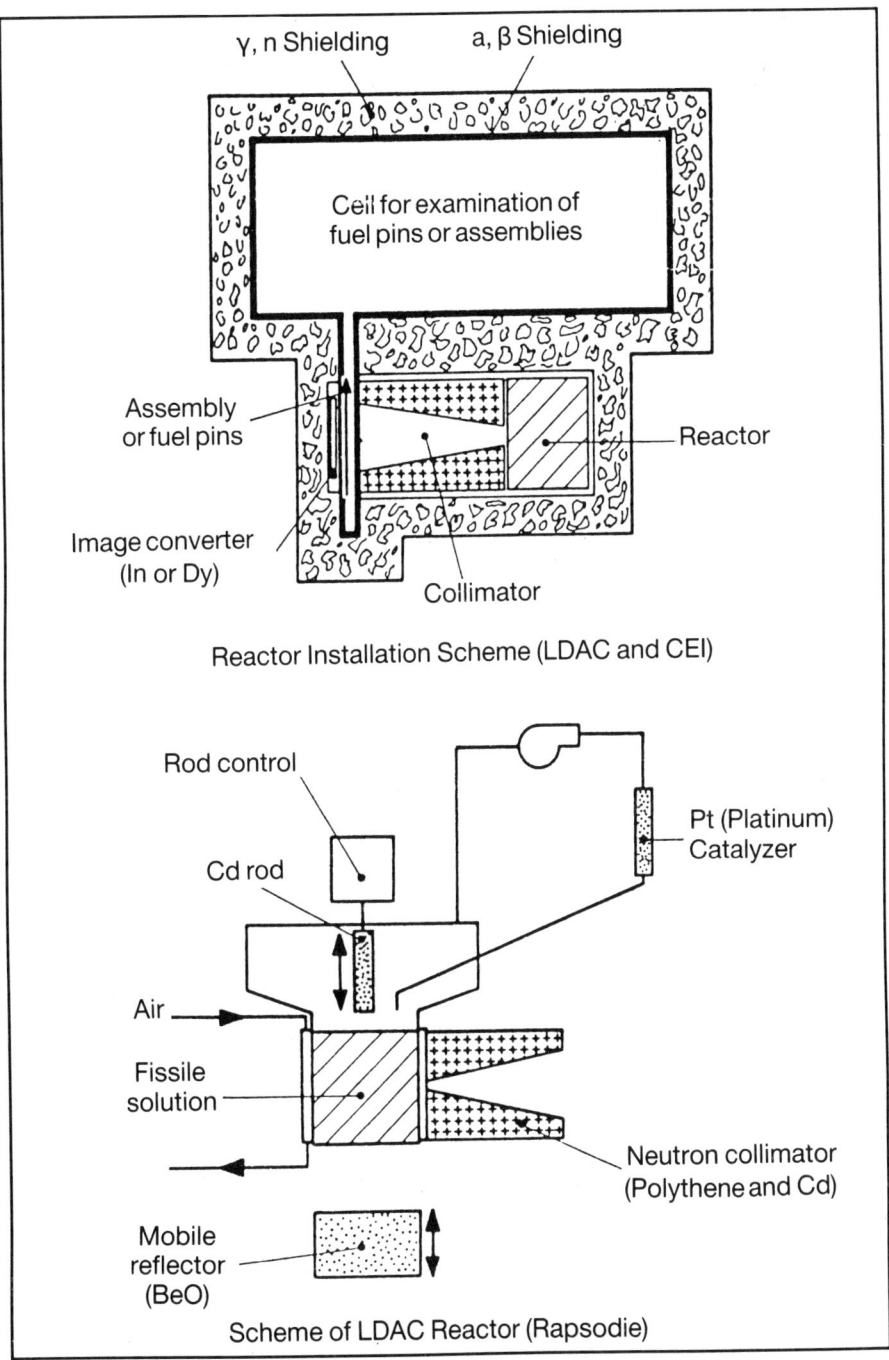

Fig. 33. *Reactor installation scheme (LDAC and CEI) (can also be found in /2/).*

4.8. Germany

In *Federal Republic of Germany* three centers operated neutron radiography facilities.

The NR facility at the *FR 2* reactor in *Karlsruhe* is described in /83, 84/ and listed also in /2/, from which fig. 34 is reproduced. It gives the general lay-out of this "dry" facility.

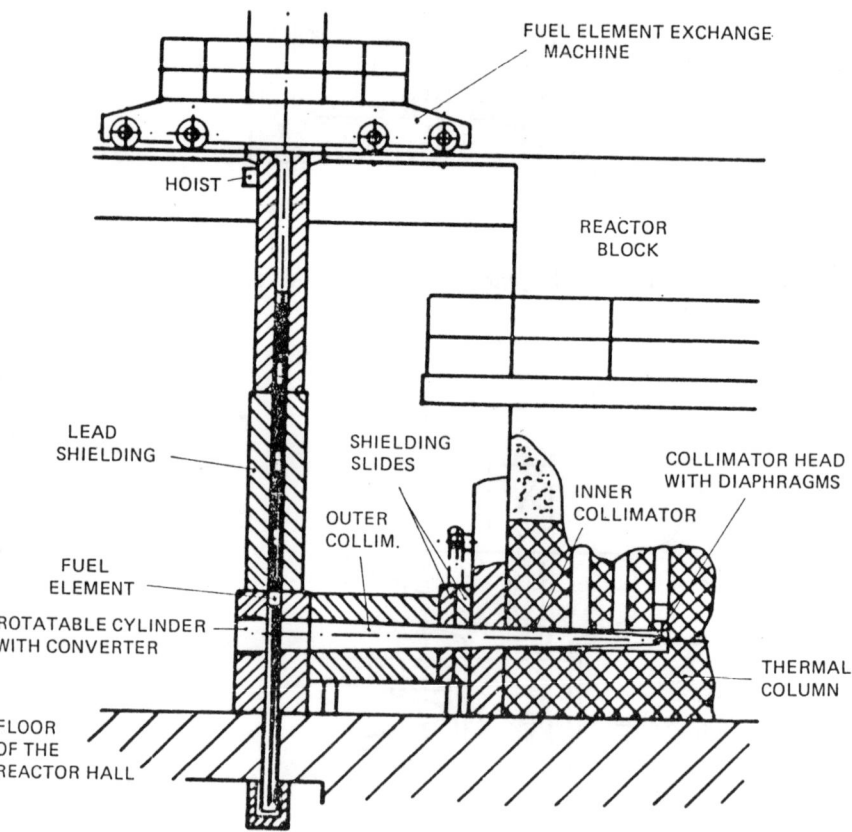

Fig. 34. Neutron radiography facility at FR2 in Karlsruhe

Some details about the collimator used at FR2 are given below. They are extracted from /84/.
The details of the construction of a divergent collimator, shown on fig. 35, are shown in fig. 36.

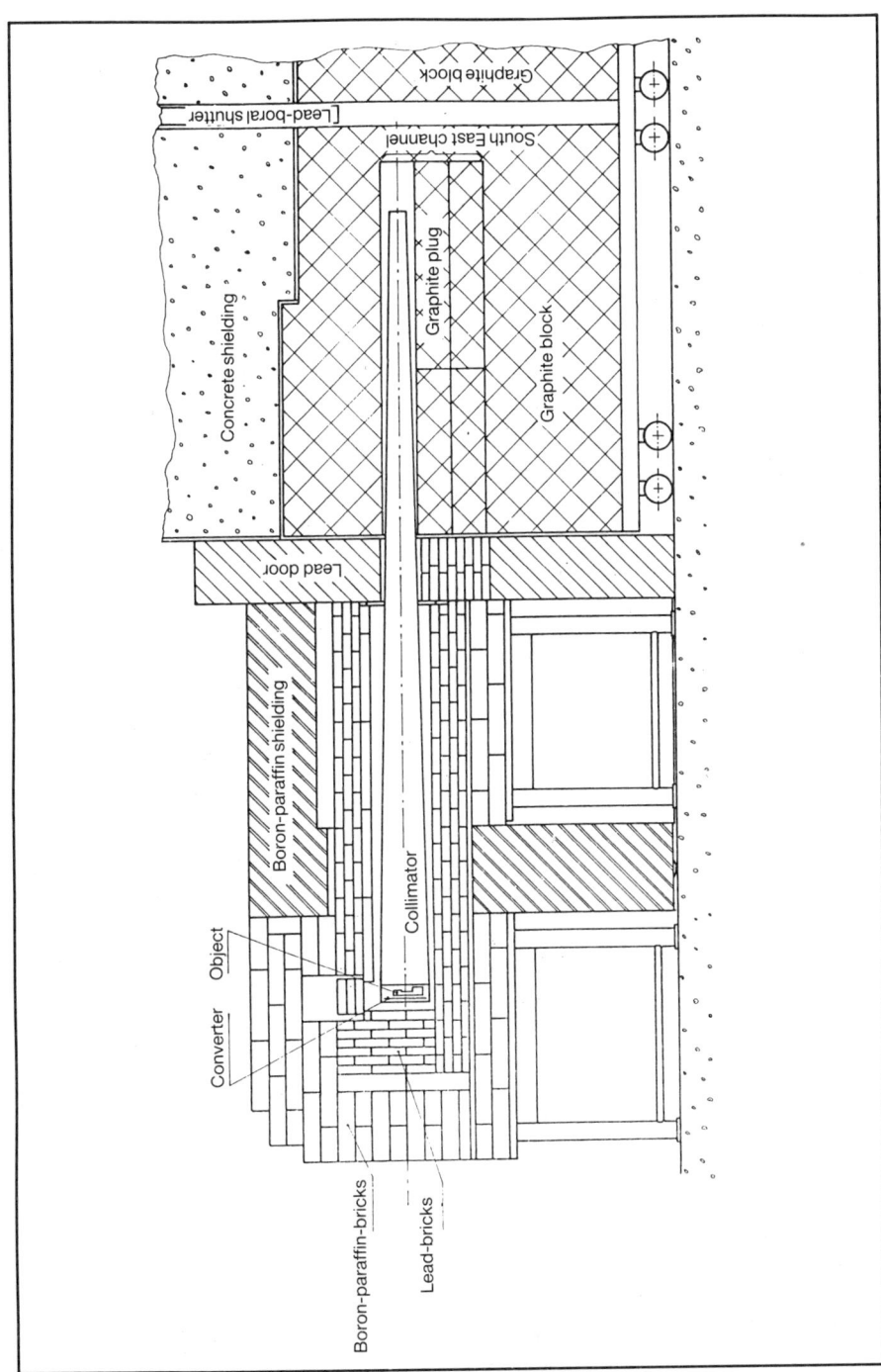

Fig. 35. Divergent collimator at FR2

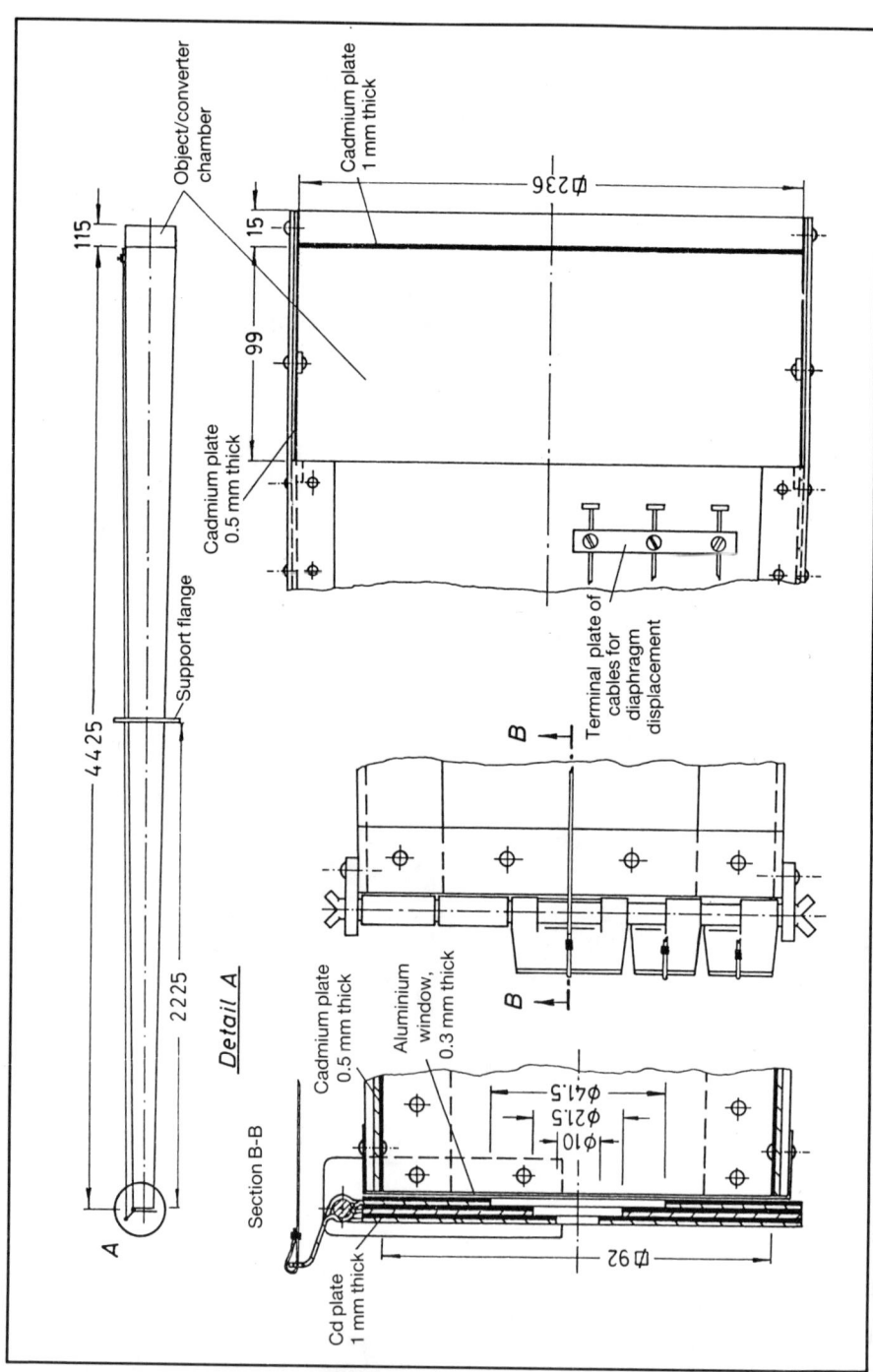

Fig. 36. Details of collimator used at FR2

The collimator is 4425 mm long, has a square cross-section with an entrance opening of 92×92 mm and an exit of 236×236 mm. By inserting different cadmium clad diaphragms at the entrance to the collimator (103.8 to 10 mm in diameter) the L/D ratio can be changed from 42.6 to 442.5. The neutron flux at the object to be radiographed will be changed from 5.5×10^6 to 6.5×10^4 n.cm^{-2}.s^{-1}.

The collimator was originally filled with air, but it was intended to fill it later with helium or argon.

As reported in /85/ studies were made at *Jülich* with the *NR-1* facility (see fig. 37) to locate it at one of the horizontal beam tubes of the FRJ-1 reactor. The beam channel of the NR-1 consists of an outer cylindrical tube in aluminium and an inner conical tube in cadmium. This inner collimator tube is held in the outer tube in adjustable holders and is 2155 mm long. The entrance of the collimator has a diameter of 50 mm and the exit diameter is 115 mm. The L/D is also 43.1.

At 10 MW power output of the reactor the thermal neutron flux at the object to be radiographed is 10^5 n.cm^{-2}.s^{-1}. A bismuth filter placed at the entrance to the collimator reduces the gamma-ray content to 2.8 R/h.

No other details about the design and construction of the NR-1 collimator are given in /85/.

At the *GKSS* nuclear center in *Geesthacht* NR facilities were established on two reactors: FRG-1 and FRG-2.

The *GENRA I* facility at the *FRG-1* reactor (5 MW) is shown on fig. 38. The description of this facility is given in /86/ from where fig. 39 is reproduced.

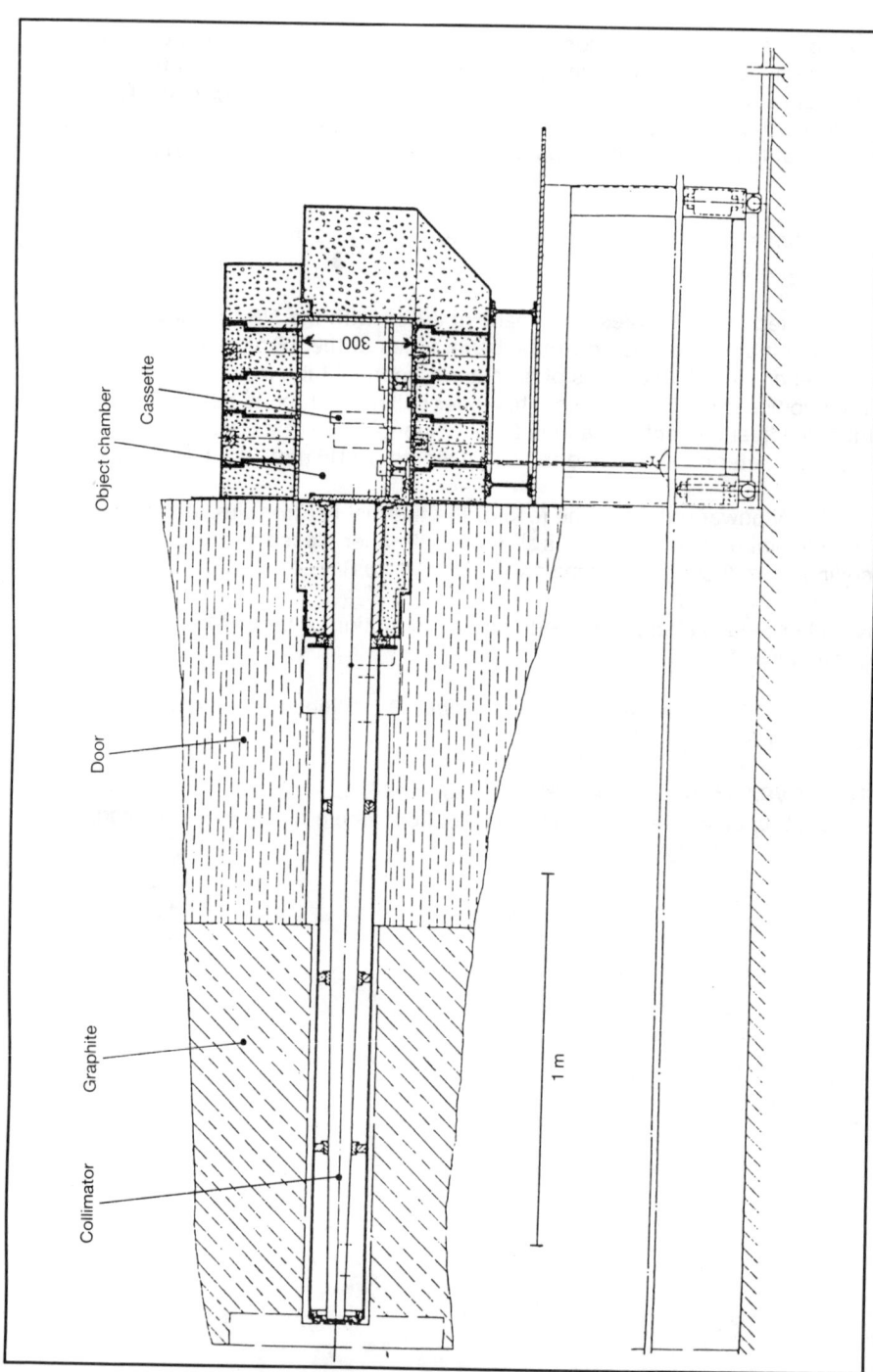

Fig. 37. NR-1 facility of the FRJ-1 reactor at Jülich

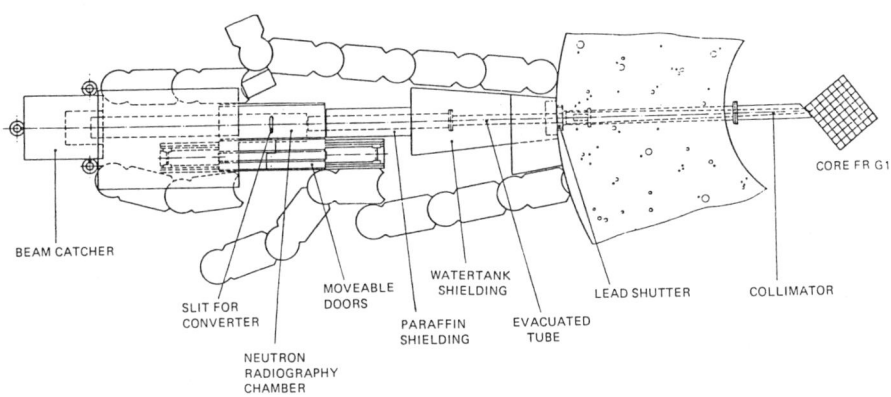

Fig. 38. GENRA I NR facilities at the FRG-1 reactor

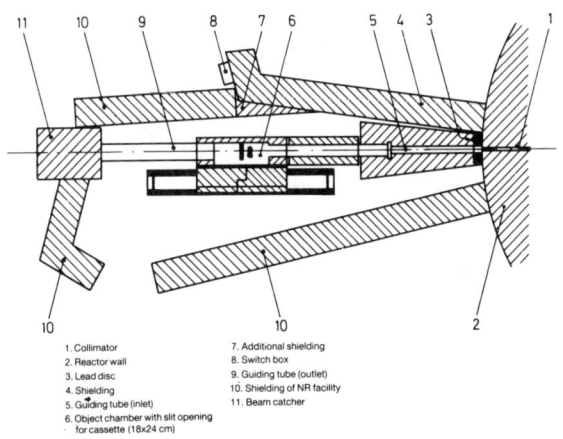

1. Collimator
2. Reactor wall
3. Lead disc
4. Shielding
5. Guiding tube (inlet)
6. Object chamber with slit opening for cassette (18x24 cm)
7. Additional shielding
8. Switch box
9. Guiding tube (outlet)
10. Shielding of NR facility
11. Beam catcher

Fig. 39. GENRA I NR facility at the FRG-1 reactor

The conical collimator is a 3.6 m long steel tube, clad from outside with a boroncarbide layer. The entrance to the collimator is of diameter of 2 cm and the exit 9 cm. The collimator is filled with water to shut down the neutron beam.

Behind the collimator a 3.9 m long evacuated aluminium tube guides the neutron beam to the exposure room. Here a neutron flux of the diameter of 18 cm has 0.8×10^7 n.cm^{-2}.s^{-1} with a collimation ratio of L/D=375.

The in-pool *GENRA II* facility at the *FRG-2* reactor (21 MW) (see fig. 40) has a rectangular collimator in the form of a pyramid.
It is 2 m long and has an entrance of 4×4 cm and an exit of 10×40 cm. The collimator is made of a 10 mm thick welded aluminium sheet. It is filled with helium. The entrance plate is 1.5 mm and the exit plate is 3 mm thick. The collimator is clad inside with a sandwich of boral, indium and cadmium.

Under the Al plate at the entrance to the collimator a sandwich of Cd and In provides the entrance diaphragm, which can easily be removed from outside. Usually a diaphragm with a diameter of 2 cm is used. Thus the L/D = 100 and the neutron flux at the object to be radiographed is 0.6×10^7 n.cm^{-2}.s^{-1}.

In the *German Democratic Republik* the *RFR* reactor at *Rossendorf* is used for neutron radiography (see fig. 41).

Here a divergent collimator is used with an entrance opening of 30 mm diameter /87/. It is composed of ten steel cylinders with an inner diameter increasing up to 40 mm. Between each of those cylinders a cadmium disk is inserted. The L/D ratio is 153 and when the power level is 10 MW a neutron flux at the object to be radiographed of 2.3×10^7 n.cm^{-2}.s^{-1} is reached, and the beam has a diameter of 9 cm.

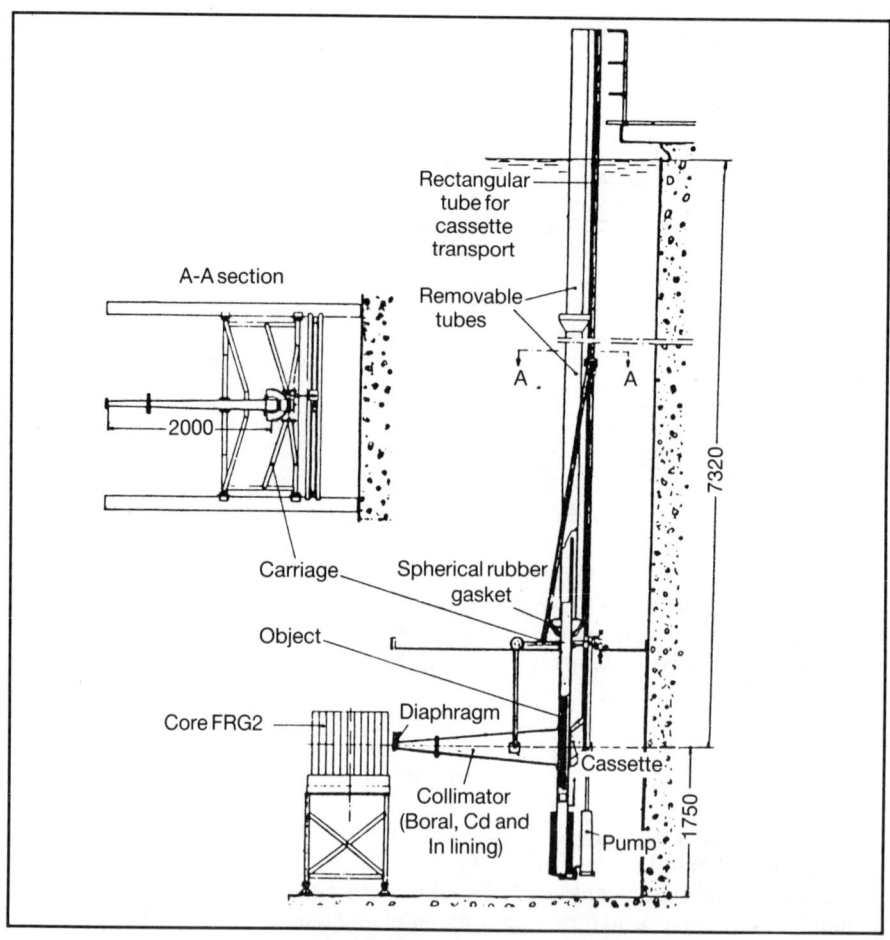

Fig. 40. GENRA II NR facility of the FRG-2 reactor

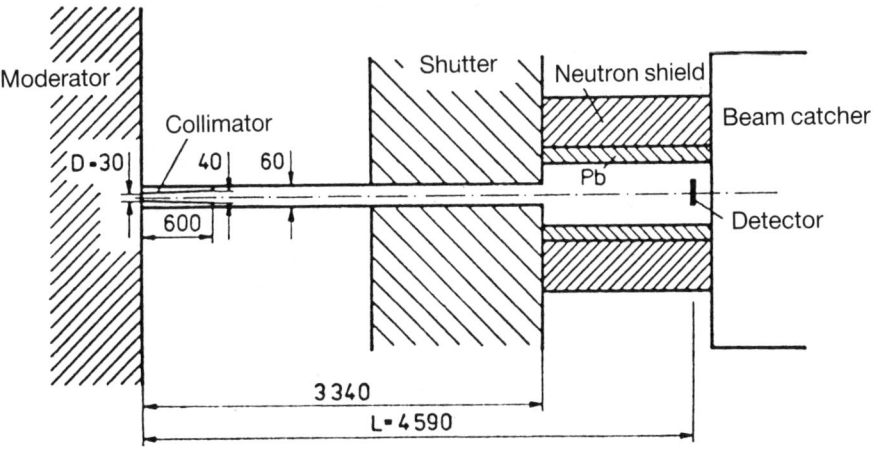

Fig. 41. NR facility at the RFR reactor at Rossendorf

4.9 India

Some information about the NR facility at the *APSARA* swimming-pool type reactor at *Bhabha Atomic Research Centre*, Bombay can be found in /88, 89, 90, 91/. At a distance of 540 cm from the core the neutron flux is 3.4×10^6 n.cm^{-2}.s^{-1} and the L/D=90.

The collimator is made of an aluminium tube (D=76 mm), 1050 mm long, lined with cadmium. This tube is set axially in a 20 cm diameter, 90 cm long wooden cylinder, which fits into the beam hole. Using a 2.5 cm aperture, the L/D = 90 can be reached.

At the *Kalpakkam* 30 kW reactor (which can be uprated to 100 kW) two beam tubes of the type shown on fig. 42 are available for neutron radiography /92/. Each beam tube is closed at the core side by a hemispherical dished end and is provided with a divergent tube collimator lined with boral, and also a beam shutter made of lead and boral.

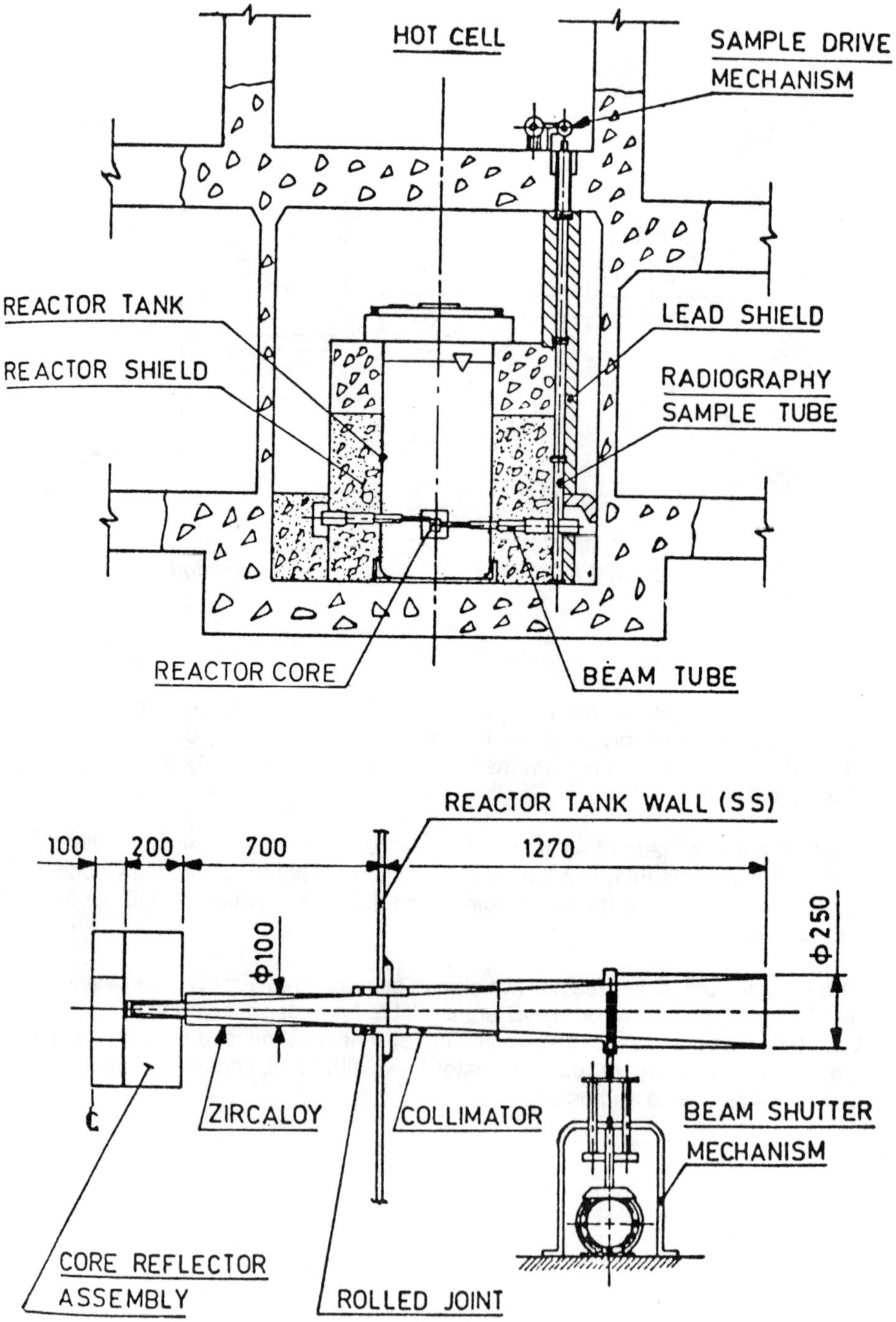

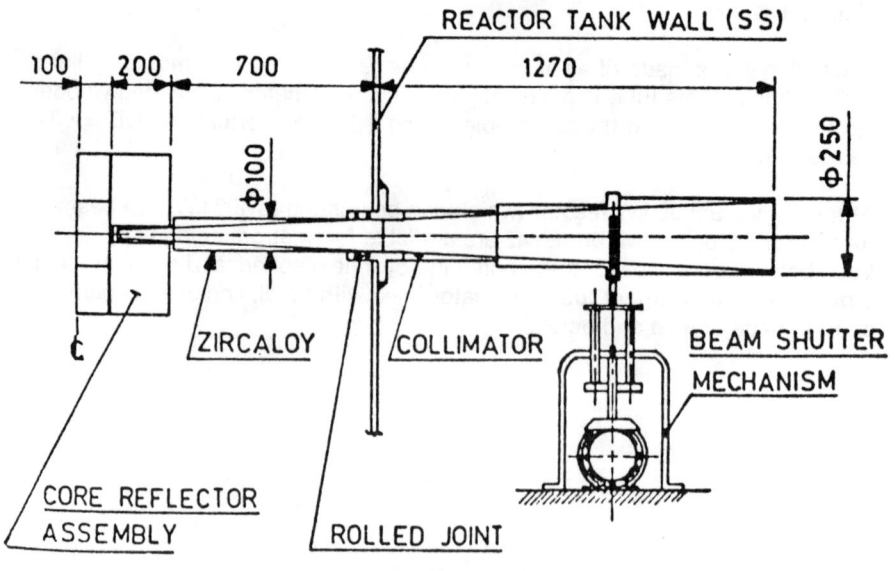

Fig. 42. General layout and beam tube details of the NR facility of Kalpakkam

4.10. Indonesia

Very scant information on the NR facility at the Research Centre for Nuclear Technique in Bandung can be found in /93/. At the *Triga Mark II* reactor a neutron flux of 5×10^7 n.cm^{-2}.s^1 at the object to be radiographed can be attained using a divergent collimator (fig. 43).

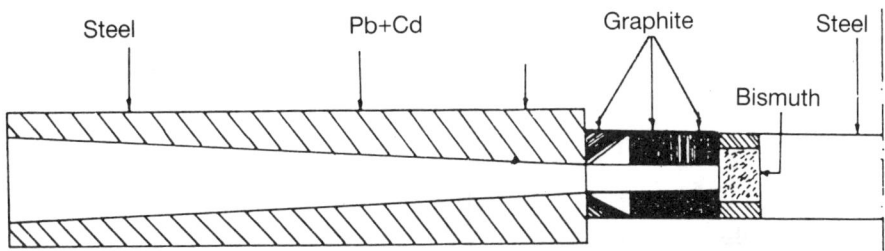

Fig. 43. Collimator of the Triga Mark II reactor at Bandung

As can be seen the neutron beam filtered by a bismuth crystal enters the lead collimator lined with cadmium.

4.11. Iran

Some information about design parameters of the NR facility at the *Teheran* nuclear research center can be found in the English abstract of /94/ (the report itself is published in Persian).

The collimator tube consists of two parts of iron and lead, 3 mm thickness of cadmium has been electro-deposited on the inside of the collimator tube. A 10 cm thick bismuth crystal is used at the core end of the collimator to reduce gamma radiation.

4.12. Iraq

The neutron radiography facility on the *IRT-2000* reactor at the Nuclear Research Institute, Tuwaitha, Baghdad is described in /95/ and is shown in fig. 44.

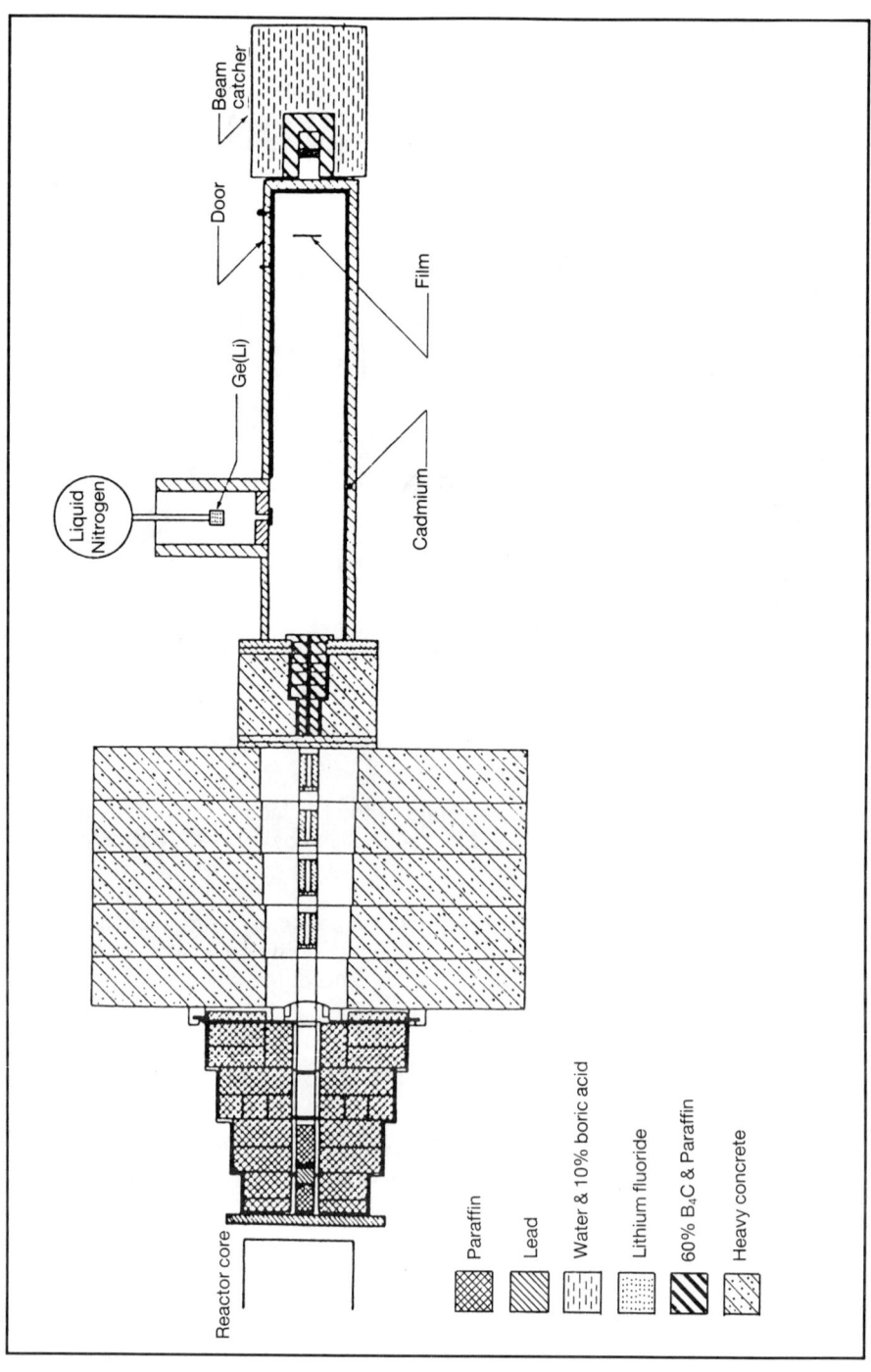

Fig. 44. NR facility at the IRT-2000 reactor in Baghdad

The IRT-2000 is a swimming-pool reactor of 2 MW. A graphite moderated thermal channel is used for NR, having an entrance opening of 15 cm. A lead plug is inserted at the beginning of the channel to reduce gamma radiation. Four converging collimators made of lithium fluoride and lead are installed along the channel within the biological shield wall, which narrows the beam to a diameter of 40 mm. A further collimator constructed of several layers of lead and borated paraffin wax and situated within the concrete shielding block further reduces the neutron beam diameter to 20 mm. The thermal neutron flux beyond this 20 mm diameter collimator was measured to be 1.5×10^8 n.cm^{-2}.s^{-1} at 2 MW reactor power. The total distance from the entrance of the thermal channel to the film is 732 cm; hence the L/D=48.8.

4.13. Israel

At the Soreq Nuclear Research Centre a beam tube of the swimming-pool type reactor has been adopted for neutron radiography /96/. A divergent type collimator is used. The cross-section of the beam tube, used at the Soreq reactor is shown on fig. 45.

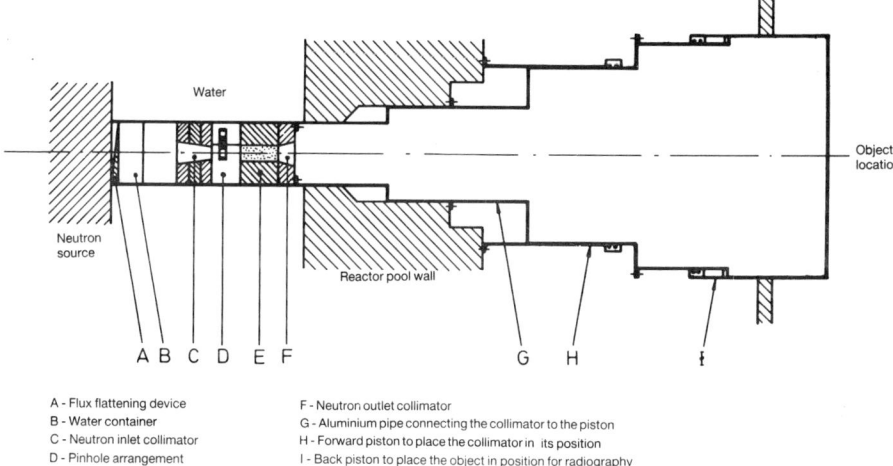

A - Flux flattening device
B - Water container
C - Neutron inlet collimator
D - Pinhole arrangement
E - Bismuth monocrystal
F - Neutron outlet collimator
G - Aluminium pipe connecting the collimator to the piston
H - Forward piston to place the collimator in its position
I - Back piston to place the object in position for radiography

Fig. 45. Cross-section of the beam tube at the Soreq reactor

The collimator proper can be divided into three parts:

(1) The neutron inlet (C) and outlet (F) are made of boral plates 6 mm thick placed 5 cm apart and filled with 92% lead and 2% cadmium alloy (see fig. 46).

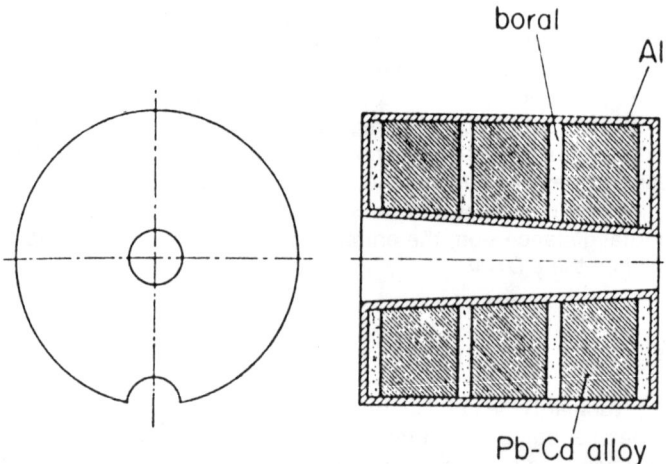

Fig. 46. *Construction of the neutron inlet collimator at Soreq*

(2) A single bismuth crystal about 15 cm long (E) attenuates gamma radiation

(3) The heart of the collimator (D) contains a disc made from a sandwich of a 0.75 mm cadmium sheet and 1 mm indium sheet between two boral plates of 6 mm each (see fig. 47)

The disc has three holes.: 8, 12 and 20 mm. The collimator is attached to an aluminium pipe, which ends as a piston outside the pool wall.

The system is designed to work with L/D ratios between 150 and 1000.

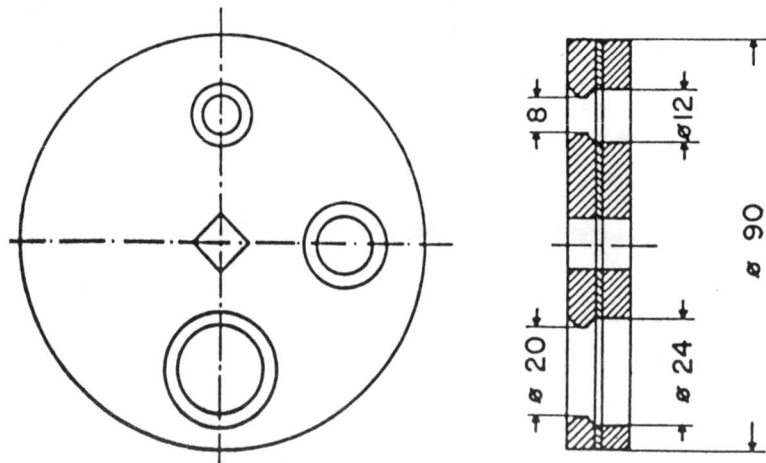

Fig. 47. *Pinhole arrangement of the collimator at Soreq*

4.14. Italy

The neutron radiography facility installed at the *1 MW Triga RC-1* reactor at CNEN–CSN Casaccia is described in /97/ (not operational at present) and shown in fig. 48.

The inside collimator is made of borated paraffin wax. It has a conical shape, a length of L = 220 cm and a minimum inner diameter of D = 4.8 cm. The external useful opening has a diameter of 12 cm. This collimator provides a neutron beam with the following parameters:

- thermal neutron flux: 5×10^7 n.cm^{-2}.s^{-1}
- cadmium ratio (measured with Au): 1.65
- collimation ratio: L/D = 50
- gamma dose-rate: 1000 R/h

Because of the high number of epithermal neutrons a Cd+In filter (3 in fig. 48) may be placed at the outlet of the collimator. All thermal neutrons are absorbed by the filter while epithermal neutrons pass through to the test area.

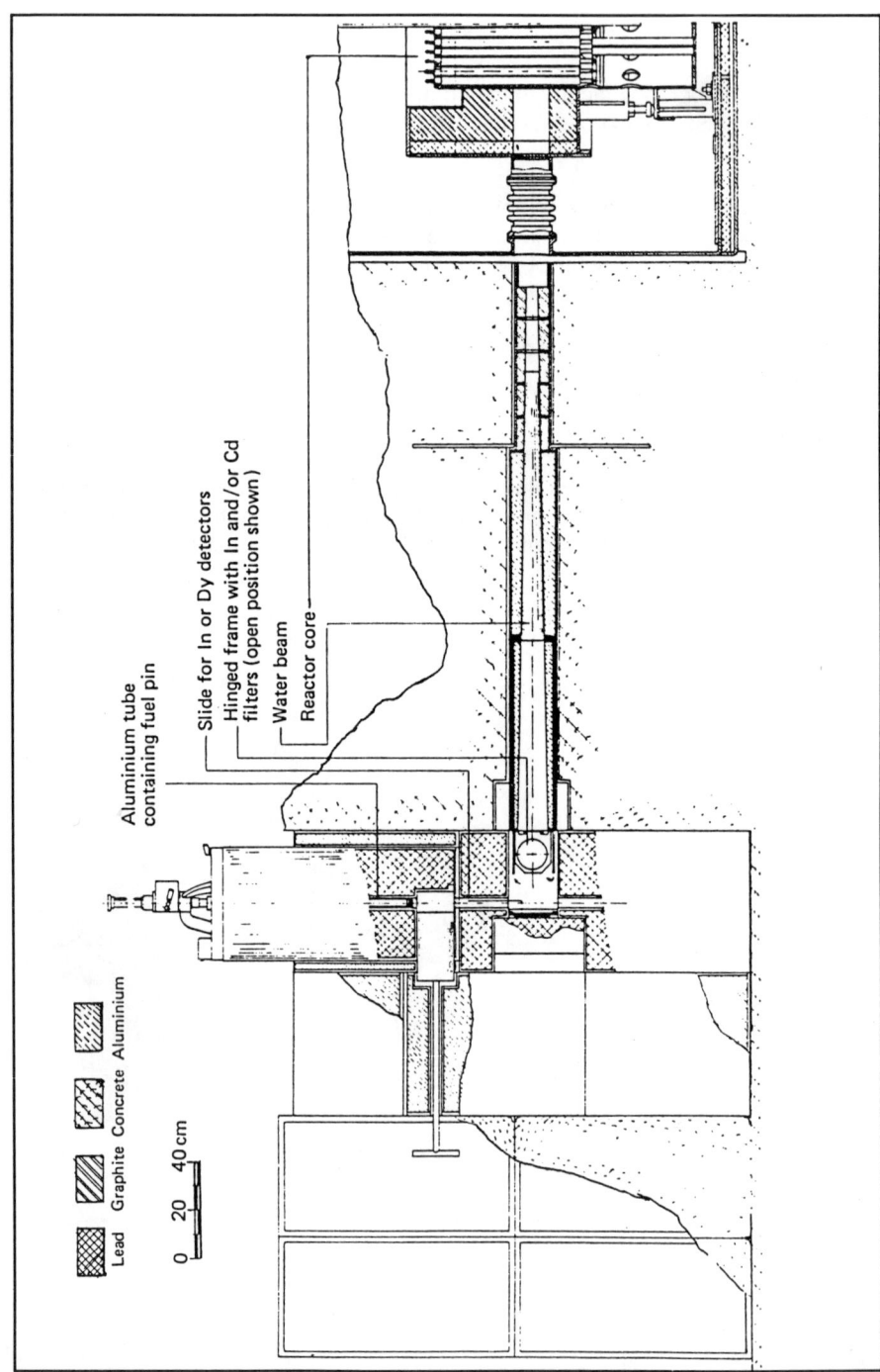

Fig. 48. NR facility at Casaccia

4.15. Japan

At the *Japan Atomic Energy Research Institute* both a research reactor and a ^{252}Cf neutron source are used for neutron radiography.

Japan research reactor No 4 (*JRR-4*), a swimming-pool type, 3.5 MW, is used for NR. Only a small amount of information is given in /98/. It consists of an 11 mm long aluminium tube, and a 4 m long straight collimator tube of 60 mm internal diameter. At the bottom of the collimator tube there is a graphite block covered by a thin aluminium plate, which contacts with the reactor core wall. Simultaneous neutron and gamma radiography, performed with the ^{252}Cf neutron source is described in /99/. The exposure device is shown in fig. 49. It uses an exchangeable divergent collimator (P) lined with cadmium. It has a 400 mm length and a diameter of 20 mm at the base, and thus L/D = 20. To improve the collimation ratio an additional 500 mm long, collimator is used and then the L/D is 45.

A 240 µg (130 m Ci)^{252}Cf source yields 5.5×10^8 n.cm^{-2}.s^{-1}.

At *Kyoto University* research reactor *KUR* (pool-type, light water, 5 MW) there are several beam holes arranged in radial and tangential geometry to the reactor core, as shown in fig. 50 /100/.

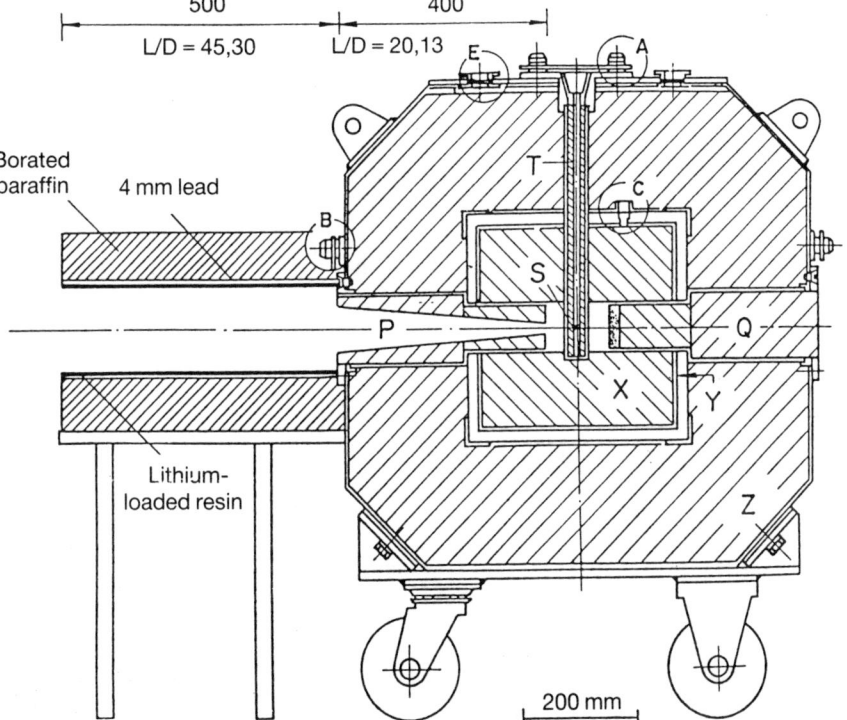

Fig. 49. Exposure device for simultaneous neutron and gamma radiography using a ^{252}Cf source

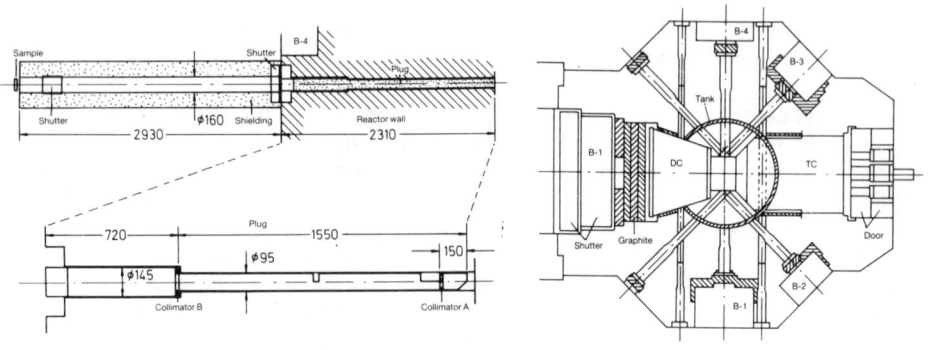

Fig. 50. NR facility at the KUR reactor at Kyoto University

With the distance to exposure position of 514 cm and the input aperture to the collimator of 10 to 3 cm the collimation ratio varies from 50 to 170 and the neutron flux at object from 1×10^6 to 7.4×10^5 n.cm^{-2}.s^{-1}.

The use of the experimental hole E-2 (see fig. 50) of the same KUR reactor is described in /101/ and shown in fig. 51.

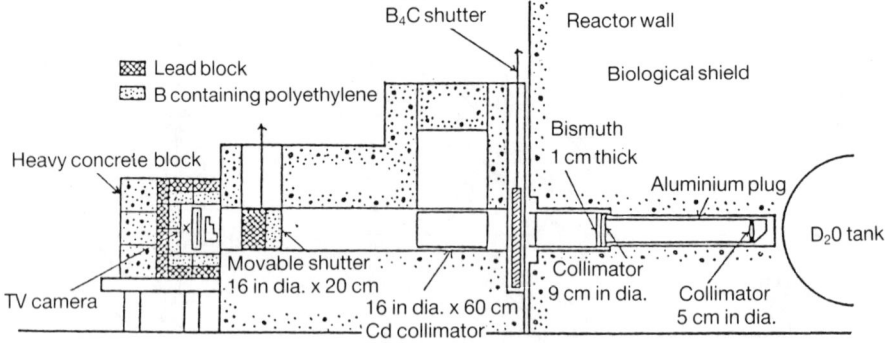

Fig. 51. NR facility attached to E-2 experimental hole of the KUR

A collimator of 5 cm diameter is inserted into the E-2 hole of 10 cm diameter to achieve better resolution. Unfortunately, neither in /100/ nor /101/ any further details about the collimator design are given.

4.16. Mexico

According to /102/ the neutron radiographic facility is located at the *Triga Mark III reactor* (light water, 1 MW) of the Nuclear Center of Mexico.

In fig. 52 the transversal view of the TW1 collimation system arrangement is shown. This includes a graphite reflector that can produce an adequate beam source and can reduce the gamma rays background arriving at the neutron image detector. It also includes a simulated neutron source obtained with a neutron absorbing shield with a small aperture and an arrangement of diaphragms to optimize the simulation conditions of the source. This small simulated neutron source is provided with a cadmium diaphragm 3 mm thick and 20 cm diameter with a 1.5 cm diameter inlet aperture at the center. Adjacent to the cadmium diaphragm, an arrangement of paraffin and lead was constructed in order to optimize the simulation conditions of the source. The angular spread of neutrons and gamma rays emerging from the simulated source are collimated by an arrangement of diaphragms of paraffin and cadmium. This complete collimation system makes the function of a divergent collimator.

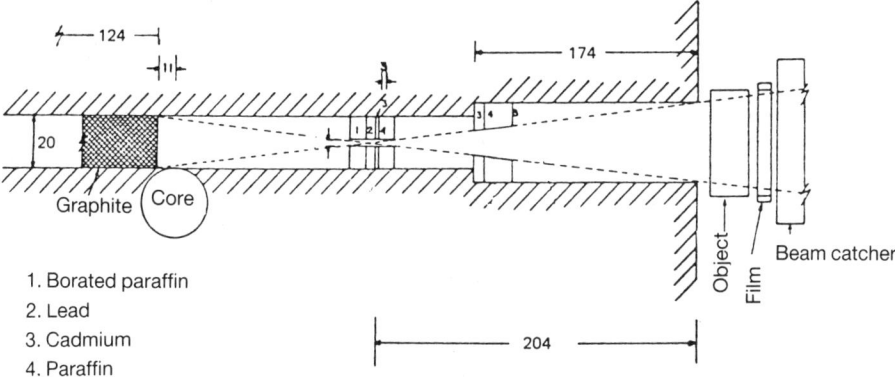

1. Borated paraffin
2. Lead
3. Cadmium
4. Paraffin

Fig. 52. Transversal view of the TW1 collimation System

4.17. The Netherlands

In the Netherlands NR facilities are located in the Netherlands Energy Research Foundation (ECN) as well as in the Joint Research Centre (JRC), Petten Establishment of the Commission of the European Communities. They are all located at Petten. At both centers nuclear reactors are used as neutron sources.

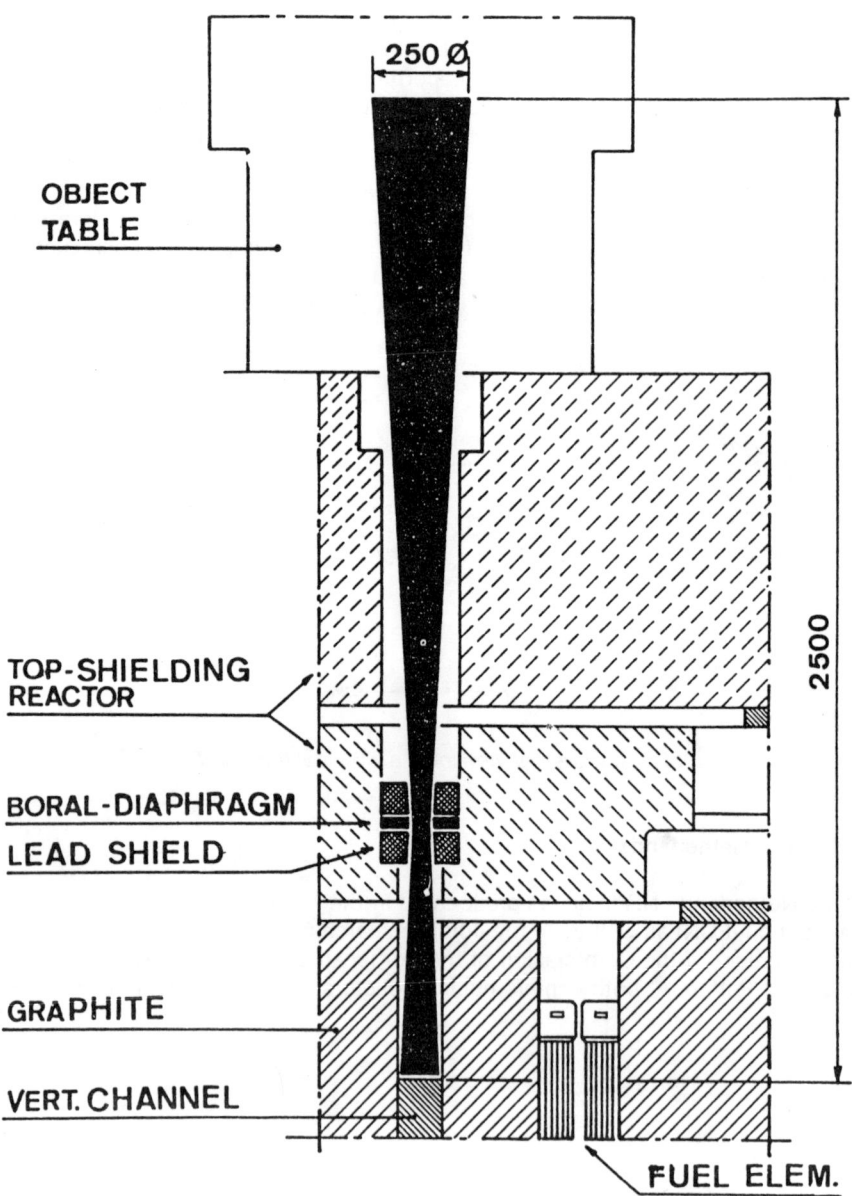

Fig. 53. Collimator System of the LFR at ECN /2/

At the *Low Flux Reactor* (LFR) at *ECN* the Argonant (Jason) type reactor, with a maximum thermal power of 10 kW is used (see fig. 53). It has a collimation ratio

L/D = 127 and a diaphragm of 15 mm made of lead rings with 2×4 mm boral. The neutron flux at the object plane is 3×10^5 n.cm^{-2}.s^{-1}.

The neutron radiography facilities at Petten *High Flux Reactor* (HFR) are described in detail in /104/. It is further reviewed in /105/ as well as in /106/ (containing all contributions of the JRC Petten to the First World Conference on Neutron Radiography /44/).

It has been designed and installed in the so-called "pool side facility", an irradiation region adjacent to the HFR core but outside the reactor vessel. Fig. 54 shows a sketch of the underwater NR installation in the pool side facility. The collimator consists of a divergent rectangular tube. Its inside is covered with boral of 6.35 mm thickness for shielding against neutrons from outside.

With the L/D = 237 the thermal neutron flux at the objects is 2×10^7 n.cm^{-2}.s^{-1}.

A former variable diaphragm (holes between 2 and 18 mm) has been replaced with a fixed inlet aperture of 8 mm diameter.

In /104/ a detailed description of neutron shielding by boral is given. As it is rather difficult to find such detail on boral shielding it is worth repeating it here.

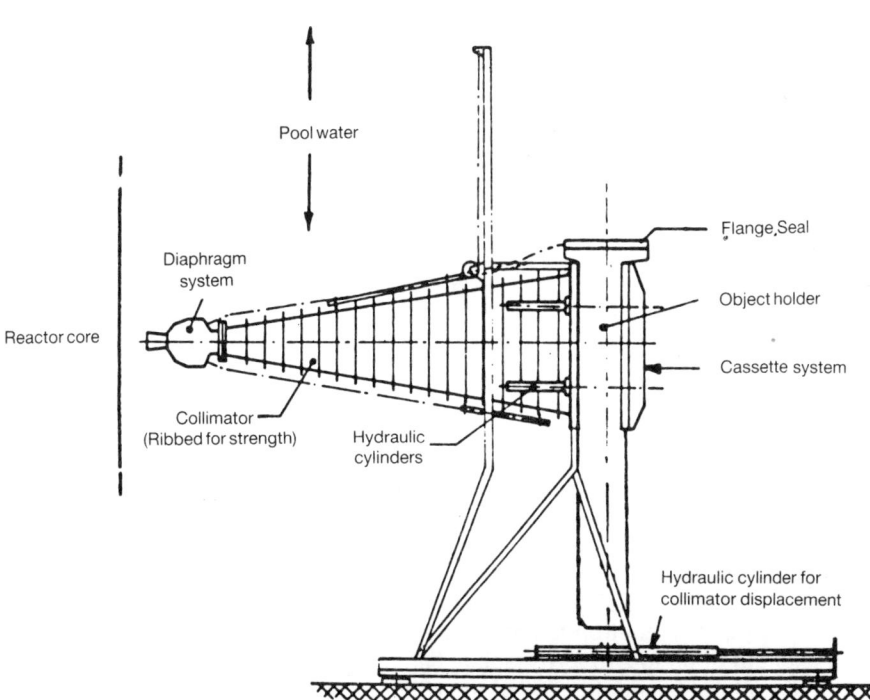

Fig. 54. Underwater NR installation in the pool side facility of the HFR, Petten.

Boral is a mixture of boron carbide particles and aluminium, made by a powder metallurgy process, sandwiched between two layers of
aluminium cladding. The boral plate for neutron shielding applied at Petten has been delivered by Brooks & Perkins (U.S.A.). Some characteristic data of it are the following:
- Total thickness 6.35 mm (1/4"), composed of a 35% B_4C dispersion in aluminium (4.23 mm, 1/16") and an aluminium cladding (1.06 mm, 1/24")
- Density: 1.68 g/cm² of plate.

Further details about the properties of boral can be found in /104/.
The collimator itself is a divergent rectangular tube of cast aluminium, consolidated by ribs (fig. 55).

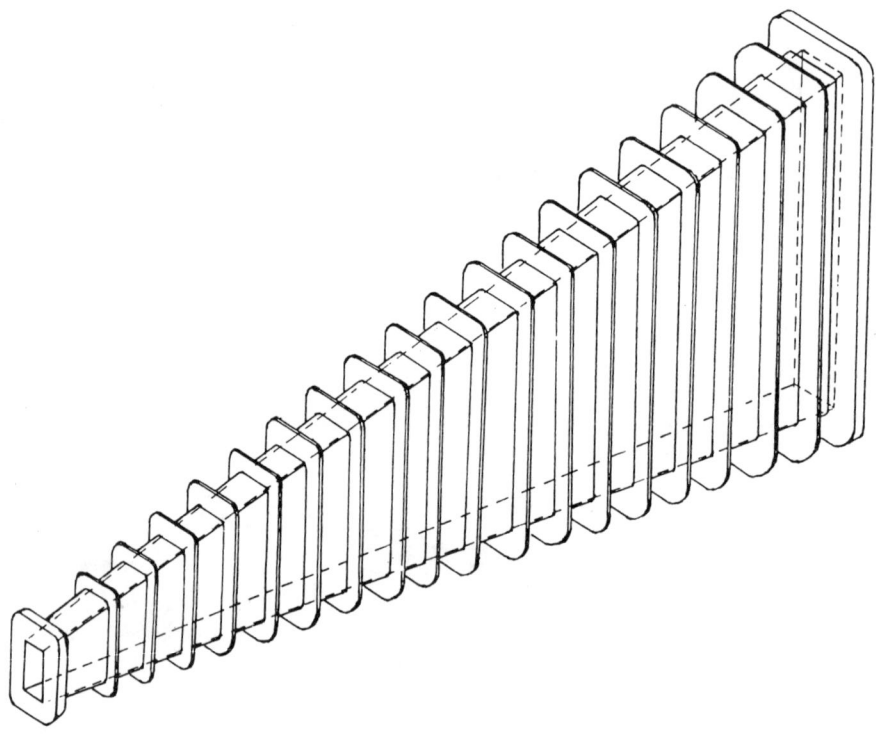

Fig. 55. Collimator of the HFR, Petten.

The inner dimensions of the front of the collimator are 3.7×8.7 cm and of the back 8.7×57 cm. The overall wall thickness is 8 mm.
To be able to perform neutron radiography of spent fuel rods from power reactors having a length of up to 2 m a NR facility has been developed at *HFR*, using a neutron beam from beamtube *HB8* of the HFR reactor /107/. This dry installation has a L/D = 500 and can deliver a neutron flux of 10^7 n.cm^{-2}.s^{-1} at the object plane. The in-pile collimator (fig. 56) can be divided into the following parts:

- aluminium front plug;
- thermal neutron inlet diaphragm (details in fig. 57);
- conical shaped collimator;
- outer tube.

Fig. 56. Collimator in the beamtube HB8 of the HFR reactor

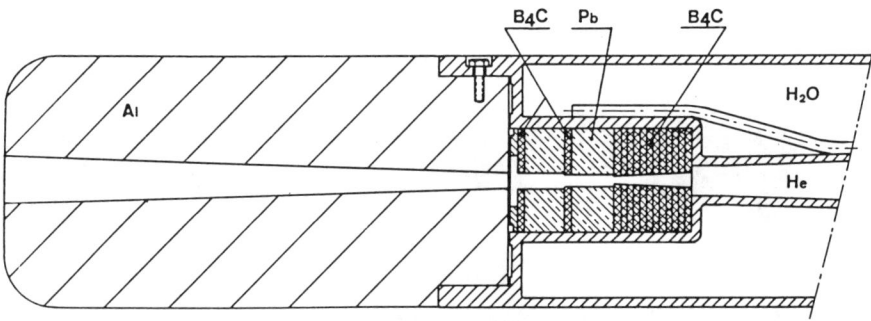

Fig. 57. Details of inlet diaphragm of collimator from fig. 56

The conical shaped collimator is lined with boral. The water around the thermal diaphragm and the conical collimator tube thermalizes the fast neutrons and thus permits easy absorption of the neutrons in the B_4C lining which might otherwise leak. The collimator is filled with helium. During 1985 the HFR reactor was renewed and upgraded by the replacement of the old reactor vessel and its peripheral equipment. At the same time a new in-pool neutron radiography facility has been provided.

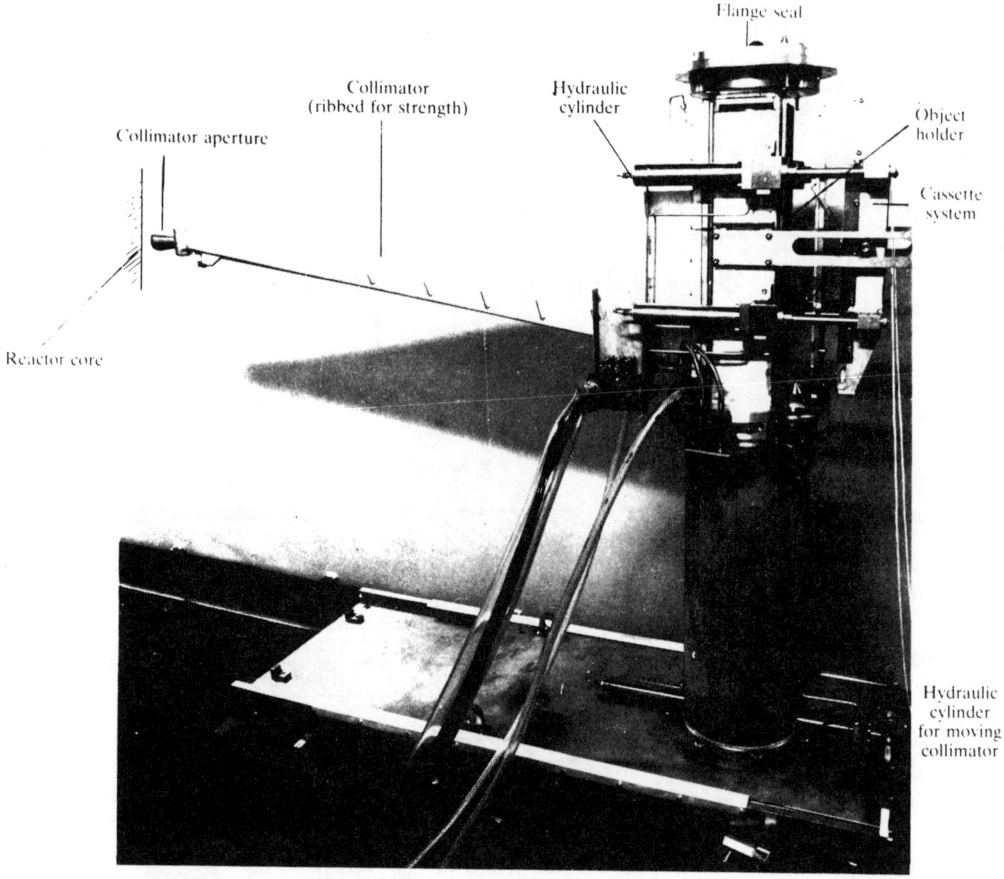

Fig. 58. New underwater neutron radiography installation at the HFR

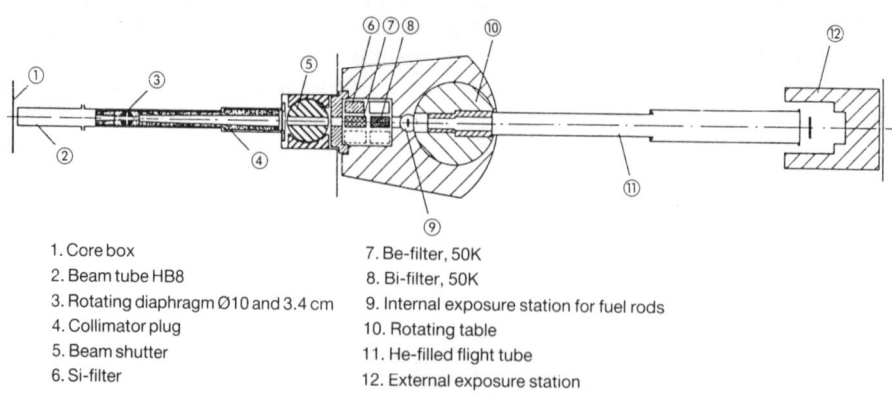

1. Core box
2. Beam tube HB8
3. Rotating diaphragm Ø10 and 3.4 cm
4. Collimator plug
5. Beam shutter
6. Si-filter
7. Be-filter, 50K
8. Bi-filter, 50K
9. Internal exposure station for fuel rods
10. Rotating table
11. He-filled flight tube
12. External exposure station

Fig. 59. Modification of the collimation system at the HB-8 dry NR facility

The HB-8 dry facility has also been redesigned /108/. It compromises the modification of the present collimation system (see fig. 59) /109/ which will be equipped with a variable aperture, a cold filter system and an extended evacuated (or helium-filled) guide tube.

4.18. Sweden

At *Studsvik* Energiteknik AB two facilities for neutron radiography are in operation. One is installed under water in the pool of the R2 reactor (50 MW MTR type) and is used for examination of nuclear fuel rods and other radioactive objects. The other is located outside the pool wall at the R2-0 reactor (1 MW) and is intended for inactive objects /43, 110/.

The general outline of the NR facility at the *R2* reactor is shown in fig. 14, which shows the main underwater parts: a collimator tube, a vertical object chamber and a duct for the holder for the image recorder.

The main data of this facility are: $D = 1$ cm; $L = 290$ cm; $L/D = 290$ cm; thermal neutron flux at object $- 2 \times 10^7$ n.cm^{-2}.s^{-1}.

The NR facility at the *R2-0* is described in /111/.

The neutron beam is extracted tangentially with respect to the core. An aluminium tank (29×25×29 cm) with heavy water is inserted in the lead window in front of the core (see fig. 60).

From inside this tank a helium-filled tube serves as a neutron guide to the collimator entrance. The collimator is installed in one of the normal beam holes in the reactor shield. The total length of the channel is 206 cm.

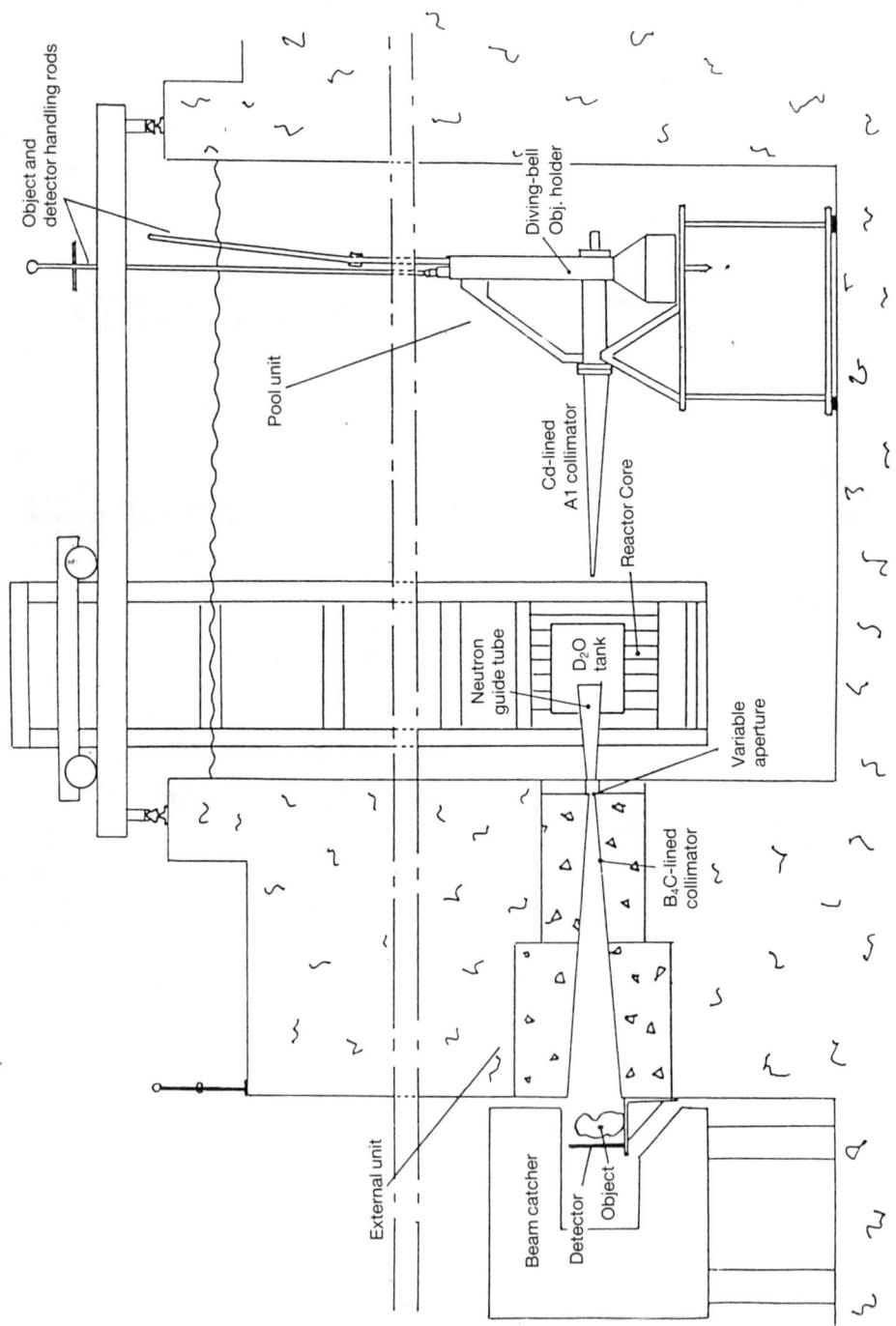

Fig. 60. NR facility at the R2-O reactor at Studsvik

A continously variable aperture is accomodated in the pocket in the lead shield (see fig. 61). It is made of two plates, each consisting of a sandwiched Al/B$_4$C/Al, the boron carbide layer having a thickness of 3 mm.

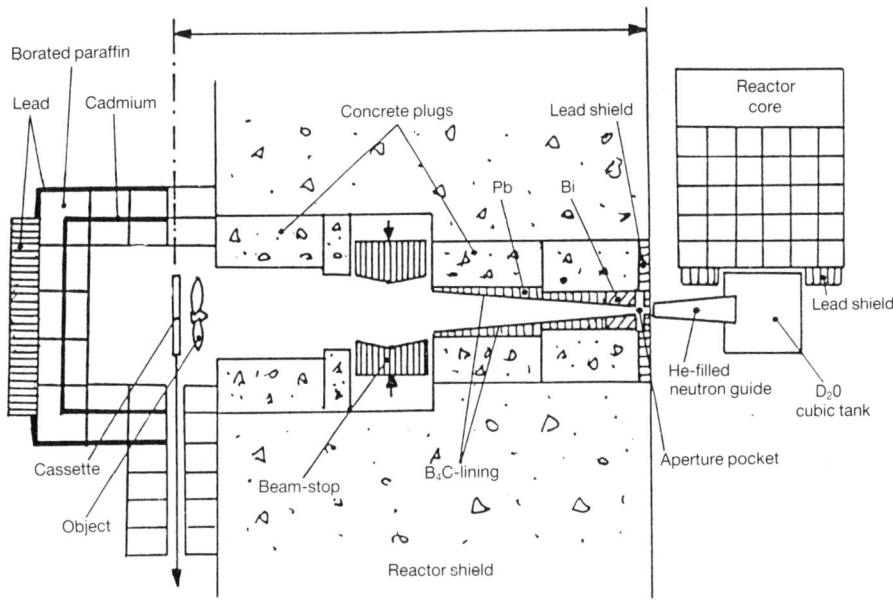

Fig. 61. Details of the NR facility shown in fig. 60

The two innermost concrete plugs are lined first with bismuth and lead and then with a 2.5 mm layer of boron carbide, leaving a conical center channel with entrance and exit holes 5×5 and 16×16 cm, respectively.

4.19. United Kingdom

The neutron radiography services in Research Reactors Division at *Harwell* are described in /112/. Only very sparse information is given on the actual design of the NR facilities. It is noted that at the *Dido* (22 MW heavy-water reactor) and *Lido* (200 kW swimming-pool reactor) divergent collimators are used.
The neutron flux at the object to be radiographed is given as 9×10^7 n.cm^{-2}.s^{-1} for Dido and 4×10^6 n.cm^{-2}.s^{-1} for the Lido reactor.

The thermal neutron beam is taken from a horizontal tube 150 mm in diameter from the *Dido* reactor. This is collimated by a boral-lined stainless steel plug followed by a floodable beam tube and pneumatically operated shutters. The flux at the cassette averages 8×10^7 n.cm^{-2}.s^{-1} with a diameter of 170 mm and L/D = 160 /113/. The thermal neutron collimator is shown in fig. 62.

The thermal neutron radiography facility at *Aldermaston* is described in /114, 115/. The main radiographic facility is situated on the H2 hole of the *Herald* light water moderated and cooled 5 MW swimming pool reactor. At the thermal neutron beam the L/D is 115 and the neutron flux $6 \times 10^5 \text{n.cm}^{-2}.\text{s}^{-1}$ at the object plane.

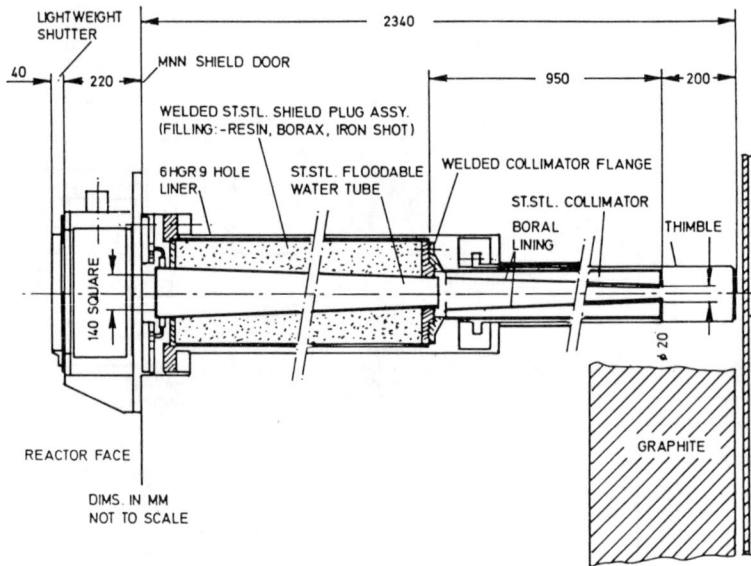

Fig. 62. Thermal neutron collimator of the Dido reactor at Harwell

The equipment for NR is shown schematically in fig. 63.

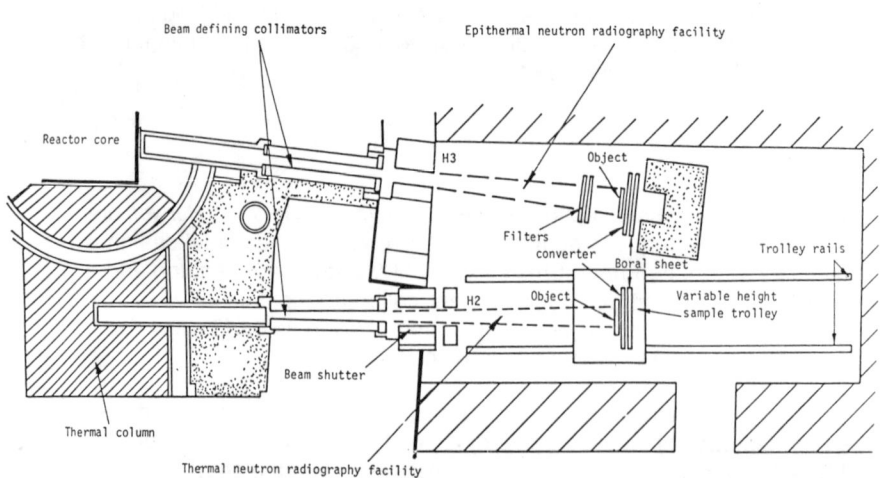

Fig. 63. Lay-out of NR facility at the Herald reactor at Aldermaston

The divergent conical collimator is 91 cm long, has an inlet aperture 40 mm, and exit aperture 62 mm in diameter.

The *Viper* pulsed reactor at Aldermaston is used for transient neutron radiography /116/. To produce a collimated thermal neutron beam for radiography the fast flux must first be thermalised by a moderator material such as polythene (see fig. 64). The side further from the reactor is covered with 0.9 mm thick cadmium to absorb

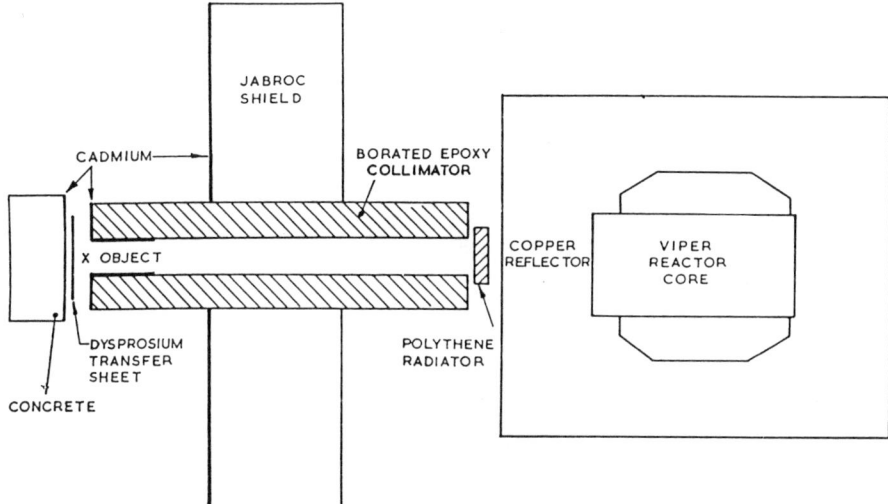

Fig. 64. NR facility at the Viper pulsed reactor at Aldermaston

thermal neutrons produced in the shield. The collimator, a 25 cm diameter cylinder 91 cm long has a 8.3 cm diameter aperture. It is cast from epoxy resin containing 60% by weight of boron trioxide and is faced with cadmium. The rear of the transfer sheet is shielded against room scattered neutrons with a 12.7 cm thick concrete block faced with cadmium.

4.20. U.S.A.

The review of the collimating systems used in various NR facilities in the U.S.A. is based mainly on the papers presented at the First World Conference on Neutron Radiography in 1981 /44/ and papers on that subject published thereafter. As the same NR facility is described several times on different occasions only those references are quoted in which some information about the collimator can be found.

The *Treat* reactor (see fig. 65), located at Argonne National Laboratory (*ANL*), Idaho Falls has been performing transient irradiation experiments related to reactor safety for more than 25 years. It also has been doing neutron radiography in the steady-stade mode for nearly 20 years. This NR facility is described in /117/ as well in /118, 119/.

The boral lined segmented, divergent collimator is 2.57 m long. It extends from the outer edge of the core cavity, through the 0.6 m diameter rotating neutron shutter, filled with 30% wt boric acid in Santowax R (a mixture of terphenyls), and into a U-shaped radiation shield.

The collimator has a vertical L/D = 30 and a horizontal L/D = 60. It has a 4.8 cm wide by 9.3 cm high opening adjacent to the core and a 14 cm wide by 48.3 cm high opening adjacent to the radiography specimens.

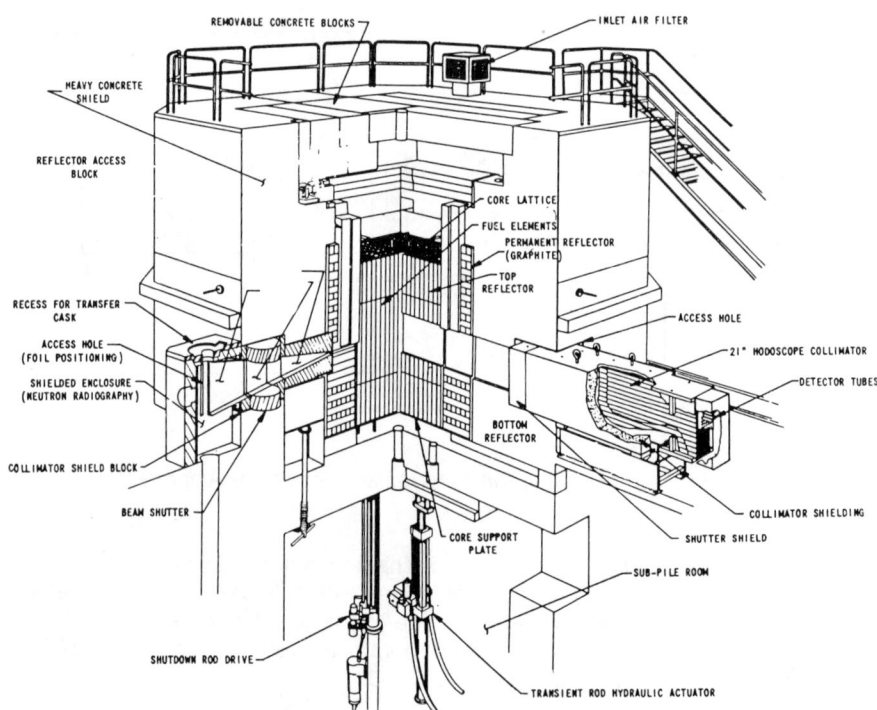

Fig. 65.   Treat reactor at ANL, Idaho Falls.

The Neutron Radiography Facility (*NRAD*) located in the Hot Fuel Examination Facility/North (HFEF/N) at *ANL* Idaho Falls is described in /120, 121/. It is also described in /122, 123, 124/. Elevation and plan views of the NRAD facility are shown in fig. 66.

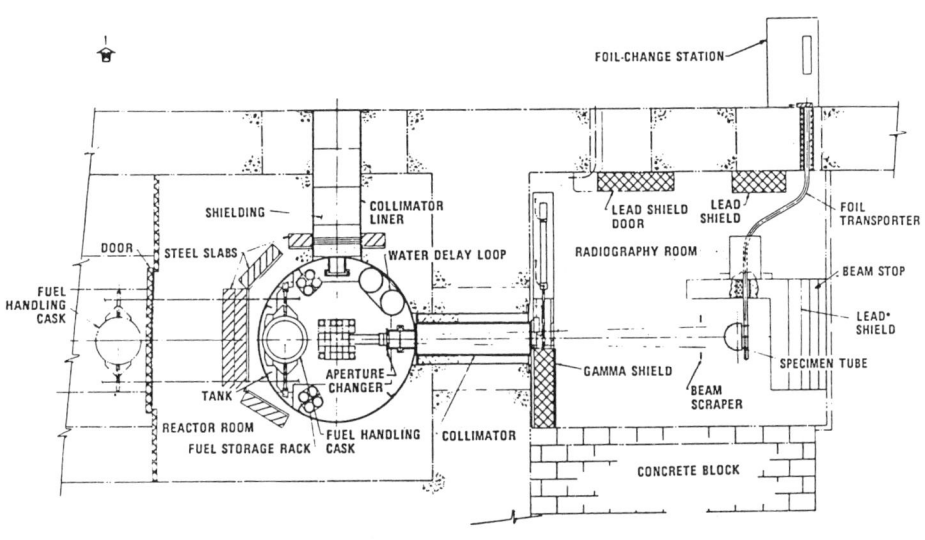
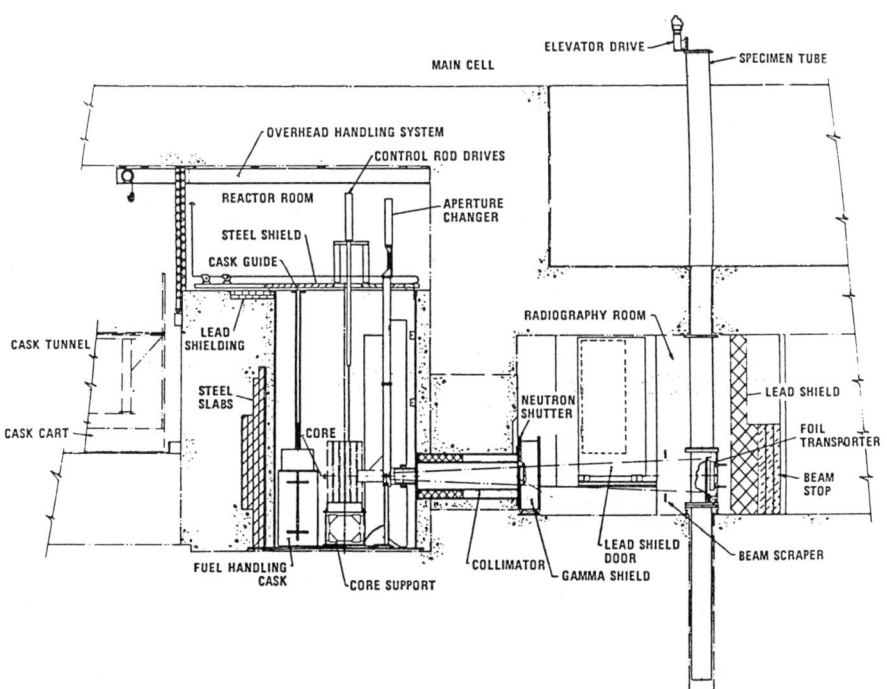

Fig. 66. NRAD facility of the ANL, Idaho Falls.

The facility has provisions for two beam tubes. The east beam tube is directly under the HFEF/N main cell, and it intersects the specimen tube extending down from the cell. The in-tank position of the beam tube consists of a changeable beam filter to tailor the beam characteristics. The changeable aperture mechanism, a boron nitride block 7.62 cm thick, is remotely operated and provides L/D ratios of 50, 125 and 300. The inlet section of the collimator is within 13 mm of the reactor fuel.

The collimator is a boral-lined tube equipped with a boral shutter. At the most commonly used L/D = 125 the flux at the imaging plane was measured to be 1.6×10^6 (0.25eV) and $1.33 \times 10^7 \text{n.cm}^{-2}.\text{s}^{-1}$.

Two papers /125, 126/ from the *U.S. Department of Energy* show only a schematic diagram of a neutron radiography test arrangement (see fig. 67) which is used for

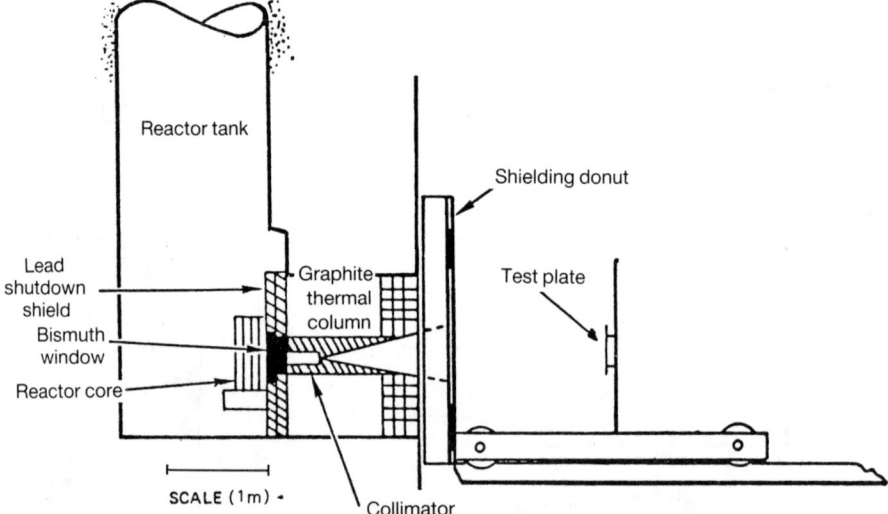

Fig. 67. NR test arrangement at Atomics International

neutron radiography of Apollo ordnance, described previously in /127/. A 1 MW pool type reactor (STIR) is used as thermal neutron source, providing a flux of $7 \times 10^5 \text{n.cm}^{-2}.\text{s}^{-1}$ at the object to be radiographed. Unfortunately, no details about the collimation system are given.

The STIR reactor is no longer used and later a 3 kW L-85 reactor was used at Atomics International. It is the same type of reactor as the DR1 reactor used at Risø for neutron radiography /62, 63/. It has vertical and horizontal neutron beams for radiography. At L/D = 64 and 80, respectively, the neutron fluxes at the object are 3.7×10^5 and $5 \times 10^5 \text{n.cm}^{-2}.\text{s}^{-1}$ /128/.

A similar reactor (*L-88*, 10 kW) designed for neutron radiography by Atomics International is described in /129/. It has a variable neutron beam collimation (see fig. 68) and a flux at the object of $10^7 \text{n.cm}^{-2}.\text{s}^{-1}$. With a L/D = 25 this can be increased to $1.6 \times 10^8 \text{n.cm}^{-2}.\text{s}^{-1}$.

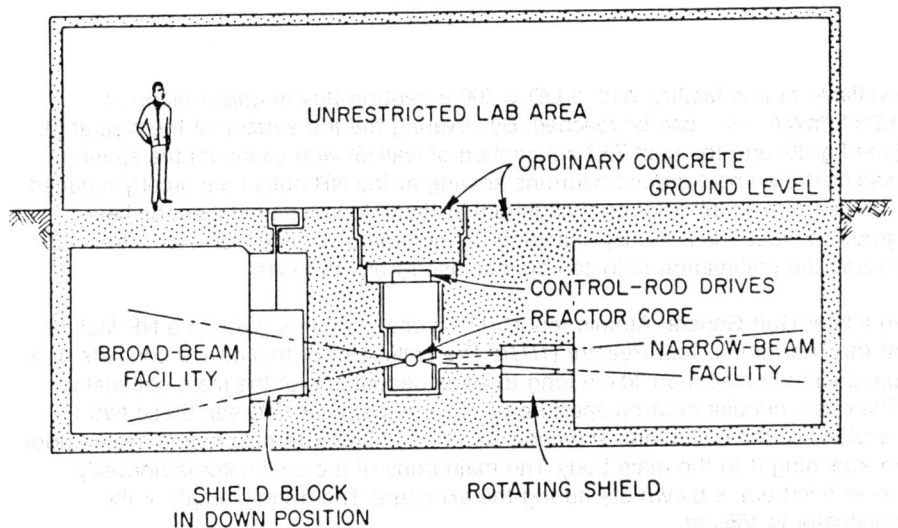

Fig. 68. NR facility at L-88 reactor of Atomics International

At Gulf *General Atomic* Co. a complete neutron radiography facility has been designed and constructed using a *Triga* reactor source /130/. This is a Mark I 250 kW graphite-reflected core also capable of producing short pulses with energy release of about 15 MWs (see fig. 69). A divergent beam collimator is

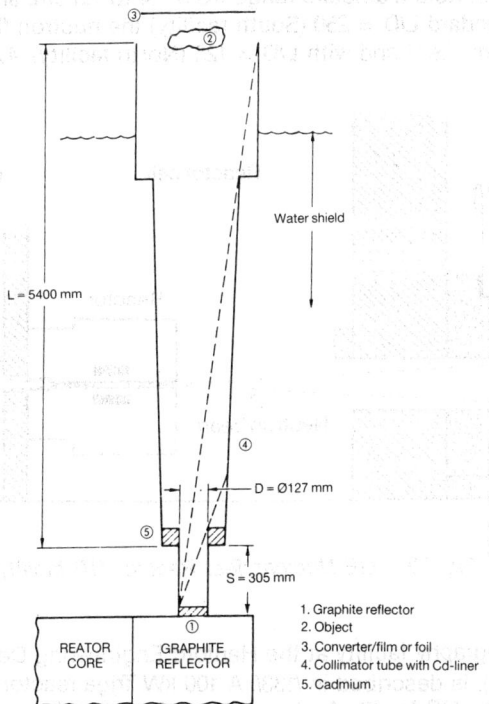

Fig. 69. Beam port of the Triga Mark I NR facility

available at this facility. With a L/D = 100 a neutron flux at object plane of $1.2 \times 10^6 \text{n.cm}^{-2}.\text{s}^{-1}$ can be reached. By covering the top surface of the step at (9) (see fig. 69) and the next 76.2 cm section of wall (5) with cadmium the spurious wall-scattered background neutrons arriving at the NR object are greatly reduced. The use of a small amount of cadmium at these locations reduces the background to less than 1%. By changing the collimator source diameter D from 12.7 to 5.58 the collimation ratio can be increased from 45 to 200.

In a later Gulf General Atomic report /131/ a description is given of a NR facility at the Thermionic Test Reactor (TITR). The collimator of this device consists of a pre-collimation section 10 cm long to allow decoupling of the main collimator. There is a circular neutron aperture of 1.27 cm diameter between these two sections of the collimator. The nose section is filled with helium and sealed prior to attaching it to the main body. The main body of the collimator is normally water filled but is blown dry during the exposure. The overall length of the collimator is 156 cm.

At General Electric Valecitos Nuclear Center the Nuclear Test Reactor (NTR) is used for neutron radiography (see fig. 70) /132/.

It is a 100 kW graphite-moderated and reflected, light-water cooled, thermal reactor. There are two NR facilities at the NTR. Each is a horizontal beam and utilizes a changeable pinhole type collimator and a graphite/polyethylene/lead source log. The pin hole diameters range from 2.4 to 7.3 cm, and the L/D from 75 to 300. With a standard L/D = 250 (South facility) the neutron flux at the object plane is $2 \times 10^6 \text{n.cm}^{-2}.\text{s}^{-1}$ and with L/D = 125 (North facility), $4 \times 10^6 \text{n.cm}^{-2}.\text{s}^{-1}$.

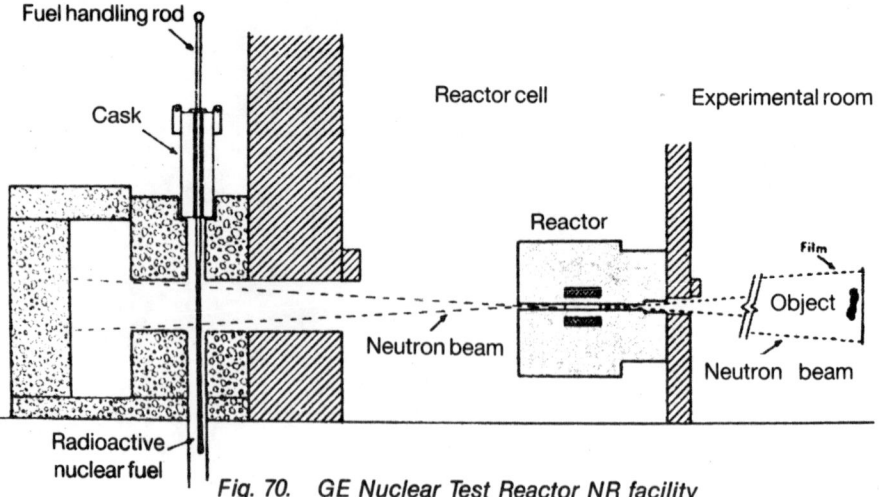

Fig. 70. GE Nuclear Test Reactor NR facility

The neutron radiography facility at the Hanford Engineering Development Laboratory (HEDL), is described in /133/. A 100 kW Triga reactor is installed at HEDL to provide an NR facility for inspection of nuclear fuel and core components.

The components and cross sections of the vertical collimator of this facility are shown in fig. 71. The aperture in the collimator consists of a cadmium-lead-indium eutectic mixture, with 0.127 mm thick gadolinium foil at the 25.4 mm square aperture. Beam shaping takes place over a 99 cm length from the aperture to top of the aperture plug. Because the required exposure is rectangular, this beam shaper, the aperture itself, and the source plane are all rectangular or square. The collimator walls are constructed in stepped sections, rather than in one continuosly diverging form. Air in the collimator is displaced by helium. The collimation ratio used is L/D = 240 and the neutron flux at the object plane is $5.8 \times 10^5 \text{n.cm}^{-2}.\text{s}^{-1}$.

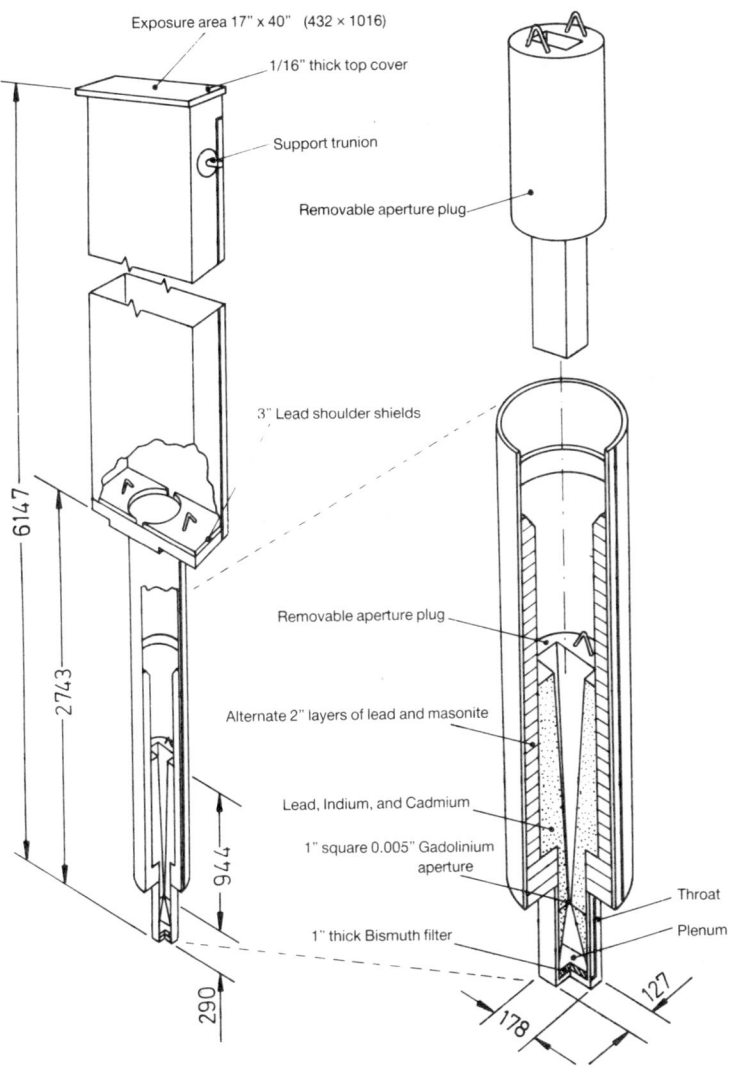

Fig. 71. Collimator for NR at HEDL Triga reactor

The design of the fuels and materials examination facility (*FMEF*) neutron radiography facility at HEDL are described in /134/. A 1 MW Mark F Triga reactor supplies neutron beams to two NR exposure facilities, operating simultaneously and independently. The two NR collimation systems are designed to operate independently, with continuously variable collimation ratios from 50 to 500. This is accomplished by using mechanical devices similar to an iris of a camera. The collimators are filled with helium. Scattered neutrons are removed from the collimators by absorption in the six boral beam scrapers that define the outer limits of each of the collimated neutron beams.

The neutron radiographic facility at the 3 MW Livermore pool-type reactor is described in /135/. In fig. 72. the plan and elevation view is given of the neutron experimental radiographic facility (*NERF*) of the *Livermore* pool-type reactor (*LPTR*). With the film plane location between 40 and 90 cm and the aperture diameter of the collimator between 12.7 and 17.2 mm the L/D varies between 190 and 300, respectively. The collimator internal shape is a divergent cone. It is portioned into three pieces to make the fabrication and installation easier. Thus the portion containing the aperture can be made small.

The sides of the collimator are encased with aluminium alloy 6061 to facilitate fabrication and handling. The shielding material is a mixture of pig lead, 99.99% pure with 3.5 to 4.0 wt% plating-type cadmium, 99.99% pure.

At the National Bureau of Standards (*NBS*) NR is performed at the Thermal Neutron Radiography Facility (*TNRF*). The 10 MW research reactor is employed for that purpose, providing a thermal neutron flux of $10^7 \text{n.cm}^{-2}.\text{s}^{-1}$ at the image plane for L/D = 40. This L/D ratio is variable from 40 to 500. A schematic diagram of the facility is given in fig. 73. /136/.

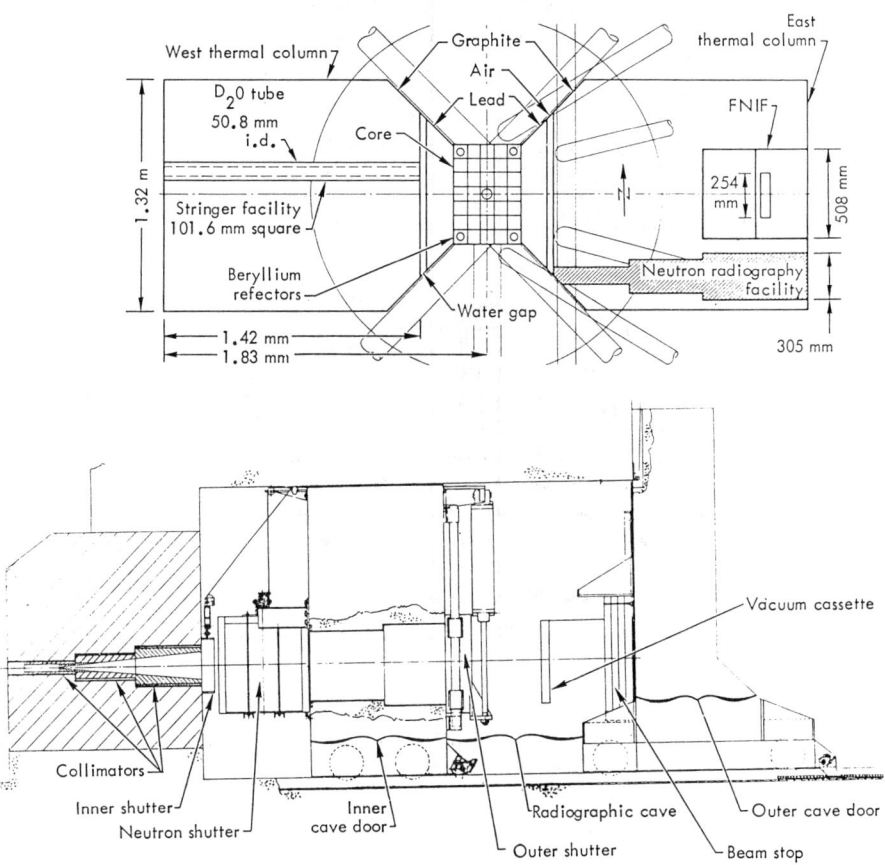

Fig. 72. Plan and elevation view of the NERF facility at the Livermore LPTR reactor

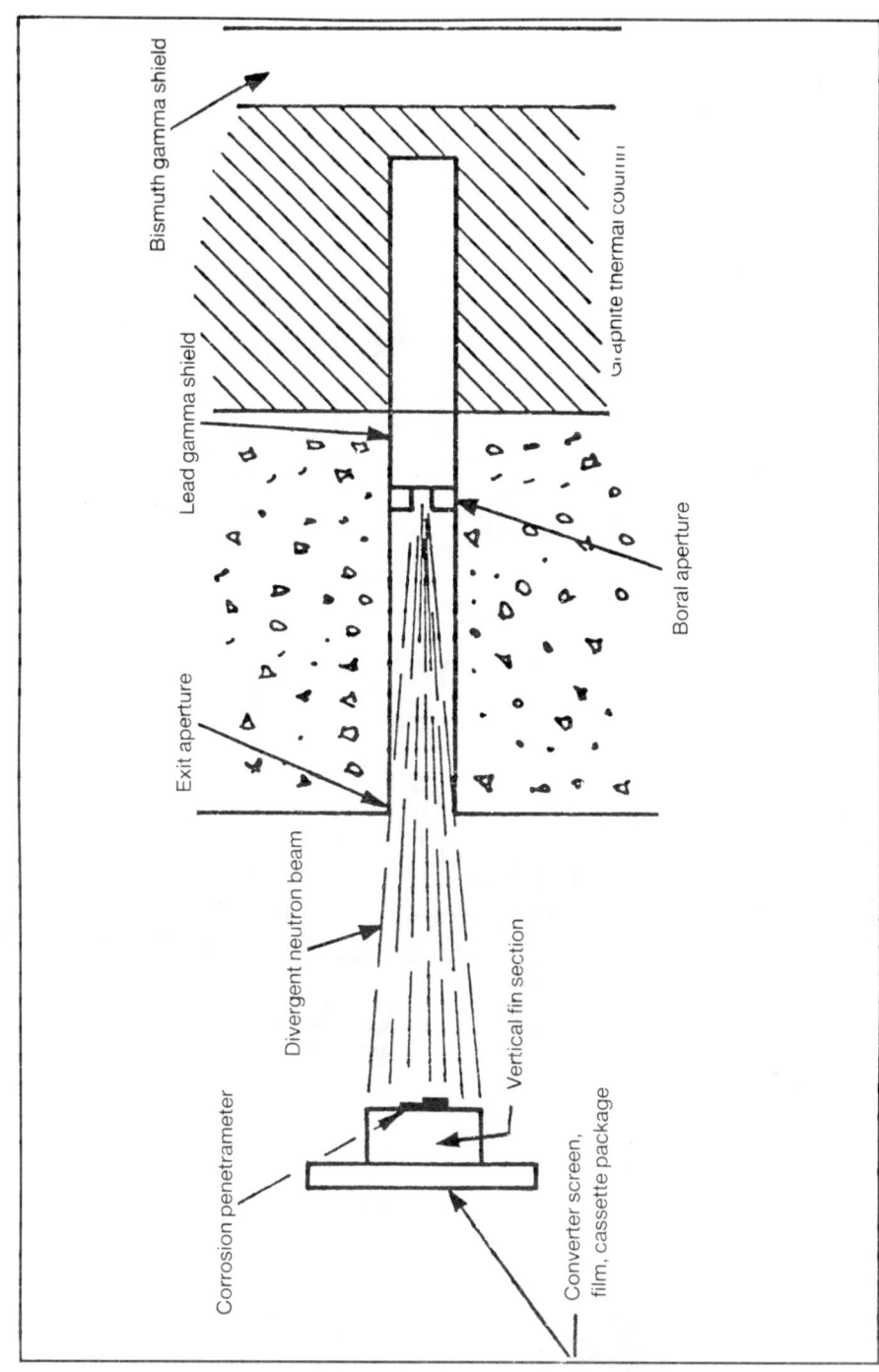

Fig. 73. Schematic diagram of the NBS TNRF facility

The neutron radiography tube of the annular core pulse reactor (*ACPR*) at Sandia National Laboratories, Albuquerque (*SNLA*) is shown in fig. 74 /137/. The tube consists of three sections.

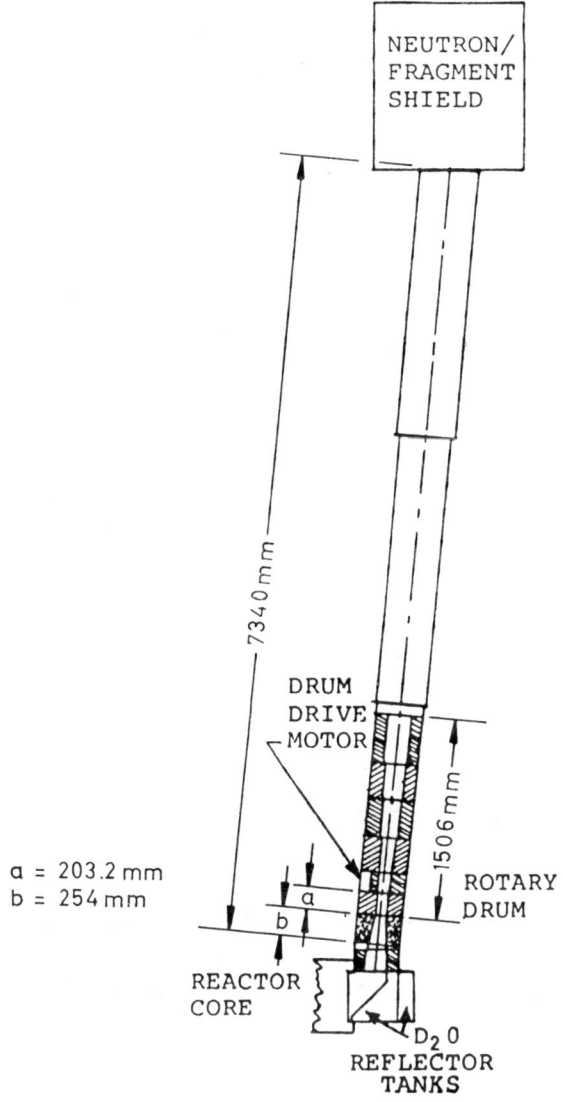

Fig. 74. Second ACPR NR tube of SNLA

A variable collimation ratio is remotely selectable through a rotatable drum, which is 25.4 cm high and has three convergent holes with square cross sections and one with a round cross section. Collimation ratios from 65 to 500 are thus possible.

The upper part of the collimator is made of four 33 cm thick parts which differ only in the central hole size and outside dimensions. The 33 cm dimension is mostly polyethylene. The top of each part is covered by a 6.35 mm stainless steel plate, a 25.4 mm lead plate, and a 1.6 mm cadmium plate. The bottom part has a tapered square hole 85.3 mm at the bottom and 118.9 mm at the top. The second, third and fourth parts have straight through square holes of 116, 146 and 176 mm respectively. The length of the tube from the aperture of the collimator to the top is 708 cm. The top plate is 560 mm square of 0.76 mm thick titanium.

Flux levels for collimation ratios from 65 to 500 vary from 1.4×10^7 to 2.93×10^5 for helium fill and from 1.05×10^7 to $6.73 \times 10^5 n.cm^{-2}.s^{-1}$ for air fill.

The NR facility at the 5 MW *University of Missouri* research reactor is built into the reactor thermal column (see fig. 75).

The collimation system is described in /138, 139/ as follows:
A lead and boron carbide matrix is selected for the collimator material because of its low cost, ease of handling, and suitability for shielding both neutron and gamma radiation. The matrix material is made by blending fine powders of lead and boron carbide in varying weight ratios with the highest boron loading in the region of the collimator aperture. Rectangular annular cans of 35.6 and 50.8 cm length are made to hold the packed material. Each can is constructed with 0.8 mm thick magnesium plate for the inner square conical collimator and with 3.18 mm thick aluminium plate for the remainder of the container. Magnesium is used because of its low activation cross section.

The collimation section (508 mm long) is a 102×102 mm parallel collimator that contains a small, removable inner can which functions as an aperture.

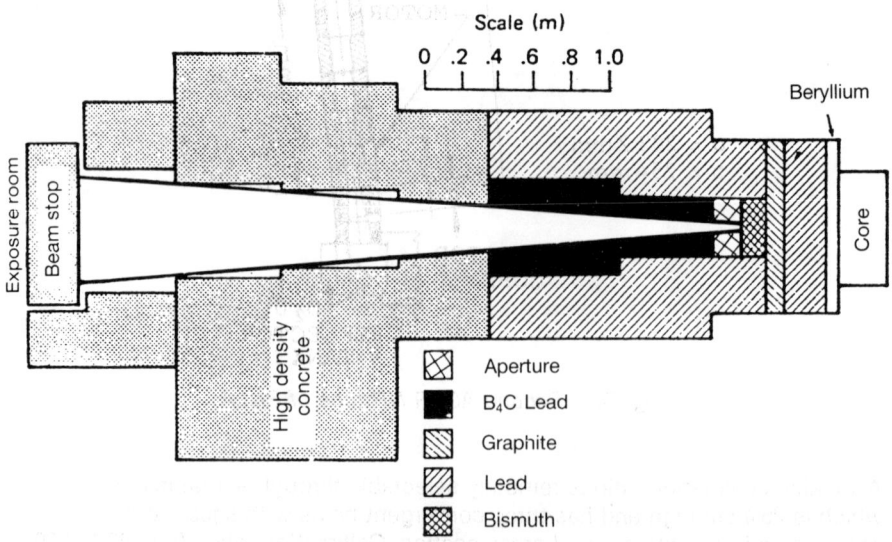

Fig. 75. Collimator of the NR facility of University of Missouri

As shown in fig. 75, neutrons leaving the core pass through 76 mm of beryllium, 330 mm of graphite, 102 mm of lead and 203 mm of bismuth before entering the variable aperture, producing collimation ratios from 28 to 200 with neutron fluxes of 10^8 to $2\times10^6 n.cm^{-2}.s^{-1}$.

The neutron radiography facility in the Oak Ridge research reactor, described in /140/ is shown in fig. 76. The collimator is a flat-sided aluminium tunnel (pyramid shaped) 175 cm long.

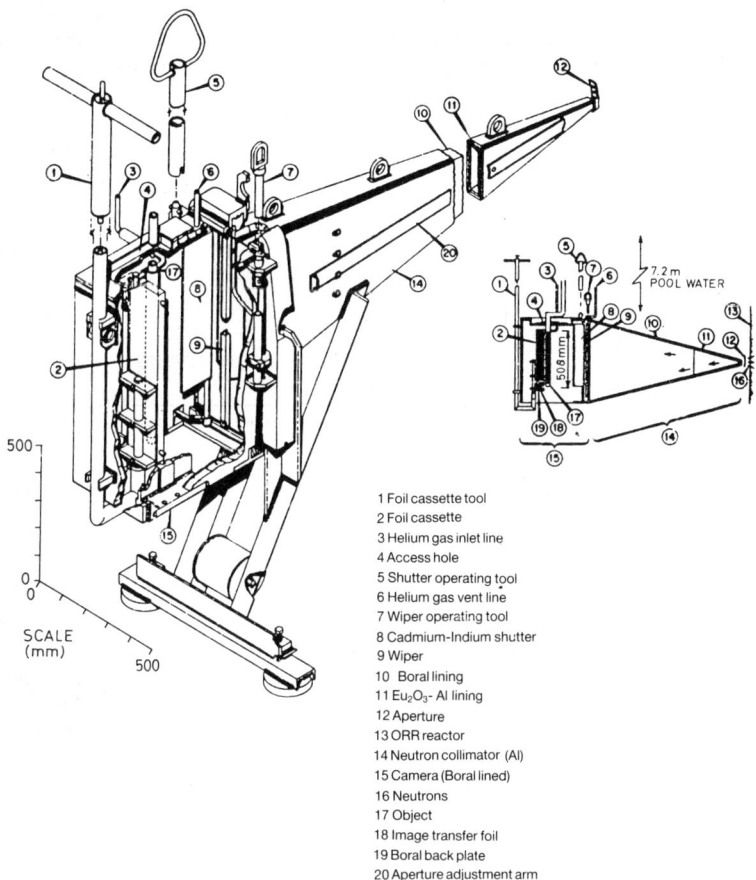

1 Foil cassette tool
2 Foil cassette
3 Helium gas inlet line
4 Access hole
5 Shutter operating tool
6 Helium gas vent line
7 Wiper operating tool
8 Cadmium-Indium shutter
9 Wiper
10 Boral lining
11 Eu_2O_3- Al lining
12 Aperture
13 ORR reactor
14 Neutron collimator (Al)
15 Camera (Boral lined)
16 Neutrons
17 Object
18 Image transfer foil
19 Boral back plate
20 Aperture adjustment arm

Fig. 76. NR facility in the Oak Ridge research reactor

It is lined with neutron absorbing materials except at the ends. The first 457 mm (from the narrow end) are lined with Eu_2O_3-aluminium dispersion 6.35 mm thick, and the remainder is lined with 6.35 mm thick boral plate. The collimator is continuously purged with dry air. The collimator has a fixed aperture of 7.6 mm square. Attached to the collimator is a curved plate with variable apertures from 5 to 3.56 mm squares or a 2 mm diameter circle. These apertures are in Eu_2O_3-aluminium dispersion material 4.75 mm thick, encased in 0.79 mm thick aluminium. The top portion of the multiaperture assembly is made of solid aluminium, and when it covers the end of the collimator the 7.62 mm square opening is effective.

The L/D ratios available with the above facility are 1000, 570, 400 and 270. The thermal flux at the object plane is about $1.7 \times 10^8 \text{n.cm}^{-2}\text{.s}^{-1}$ with the 7.62 mm aperture.

At the Nuclear Science Center of *Texas A & M University* the thermal neutron facility (see fig. 77) can give a flux of $10^7 \text{n.cm}^{-2}\text{.s}^{-1}$ for a reactor power level of 1 MW /141/. The beam is defined by a convergent-divergent collimator which is

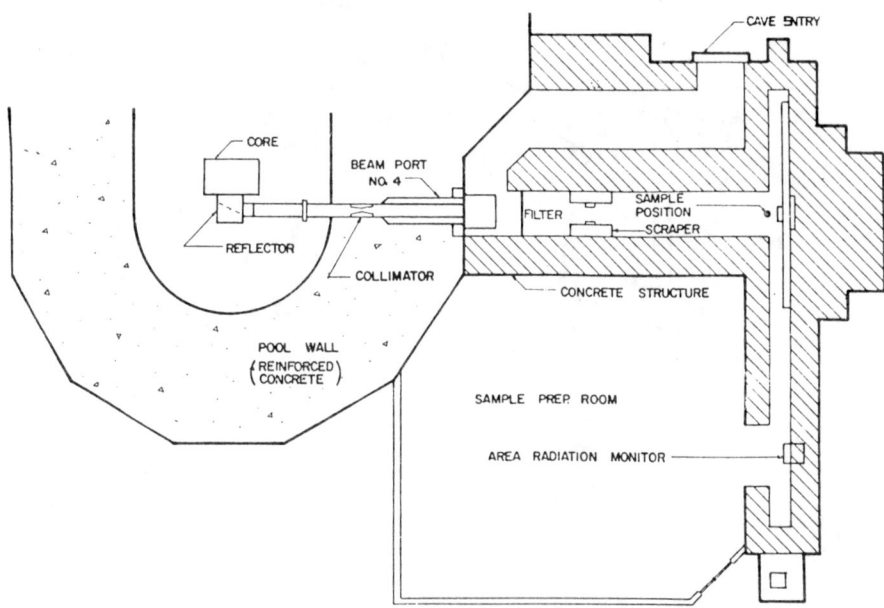

Fig. 77. NR facility at Texas A & M University

cast from polyester resin mixed with powdered lead and B_4C. The standard aperture yields on L/D = 150. This can be changed manually to 400.

4.21. U.S.S.R.

All the NR facilities described in /142, 143/ use slow neutrons for radiography (NR-11R, RBT-6, NR-21R, NR-31R and NR-32R) and therefore will not be reviewed here. Besides, no details are given about the design and construction of the collimators.

4.22. Yugoslavia

At 250 kW Triga Mark II reactor in *Ljubljana* two permanent NR facilities are in operation /144/. The thermal column NR facility (see fig. 78) is used for inspection of large objects, while the vertical beam tube is intended primarily for examination of small samples.

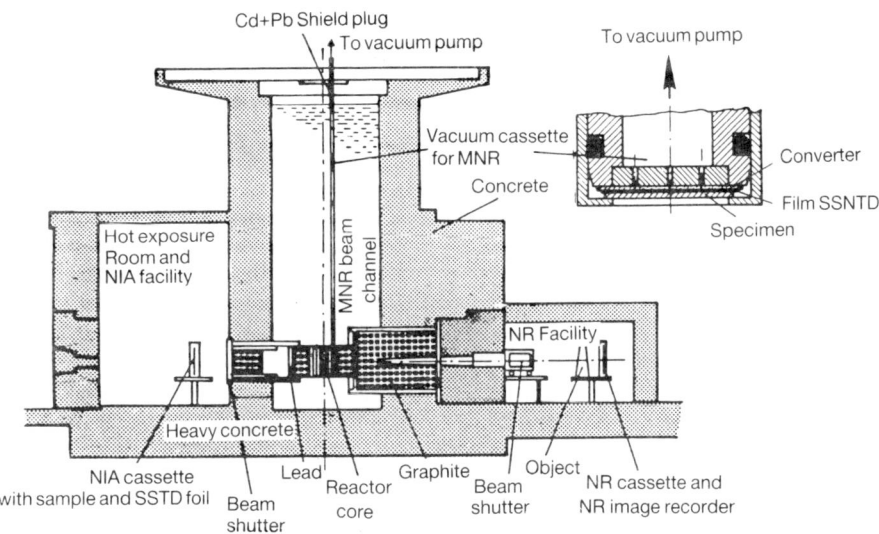

Fig. 78. NR facility at Triga Mark II reactor in Ljubljana

In the experimental channel of the thermal column the graphite scattering plug is replaced by a 2 cm long Cd lined conical collimator. The diameter of the basic aperture is 5 cm. A 10 cm thick Bi filter can be inserted. The collimator is gamma ray shielded with 25 cm thick lead rings.

At the conical collimator a neutron flux of $2.5 \times 10^5 \text{n.cm}^{-2}.\text{s}^{-1}$ is available with a L/D = 125.

4.23. Final remarks

As mentioned in 1 above the description of collimators used in different NR facilities is given without regard to wether or not those facilities are still operating.

It must also be stressed that for practical reasons it was not possible to review all the papers published on neutron radiography as the purpose of this report is to give information only on collimator design and construction. As mentioned in 1 above such information is rather sparse and therefore in many instances in the review given above some essential information may be lacking because it is not contained in the papers under review.

Of course, it is possible that some publications containing information on collimators were not reviewed above. Neverthless it is hoped that most of the important NR facilities have been described.

5. SUMMARY OF DESIGN DATA

In the previous 3 chapters collimator design data were collected from 144 publications in which such data could be found. In order to facilitate the reader's task of finding the relevant information on collimator design, a summary is made in which the most important data will be collected and reference will be made to those publications in which detailed information about collimator design can be found.

Those data which are reviewed below pertain to the following properties of the collimator:
1) Geometric shape.
2) Materials of the walls and their lining.
3) Filling of the collimator.
4) Shutters and diaphragms at the ends of the collimator.
5) Gamma-ray filters.
6) Seals for in-pool collimators.

While listing the data for the above items the proper number of the reference (listed at the end of this report) is given as well as that of the illustration reproduced from such reference, if available.

5.1. Geometric shape of the collimator

As mentioned at the beginning of this report only divergent collimators are described here.
The most commonly used physical form of the collimator is a truncated cone or pyramid.

In many descriptions of the collimators their physical form is not mentioned. So reference is made here only to those collimator designs where their physical form is explicitly mentioned.
Conical (truncated cone) collimators are used in the earlier NR facilities. Their description can be found in /5, 19, 29, 75, 85, 86, 97, 107, 135, 144/. They are represented in figs. 3, 9, 28, 37, 48, 56, 78 of this report.

A truncated *pyramid*, either with a square or rectangular cross section, is most commonly used in the design of a collimator. Reports describing such a design as, e.g. /69/ (fig. 24); /78, 79/ or /81/ (fig. 32, 33) do not always give details about the cross section of the pyramid.
The *square* collimator cross section is sometimes used, as described, e.g. in /83, 84/ (fig. 36) or /133/ (fig. 71).

However, as many objects to be neutron radiographed are elongated (such as, e.g. nuclear fuel) and the films used in NR are *rectangular*, this cross section is most commonly used for collimators. A description of this design can be found in /36, 49, 52/ (fig. 15); /54, 55, 56/ (fig. 17); /77/, /86/ (fig. 40); /104, 105, 106/ (figs. 54, 55, 58) and /140/ (fig. 76).

Some collimators, or parts thereof, have a *divergent-convergent* shape. Such a design is described in /69/ and shown in figs. 24 and 25. Here a convergent precollimator is added to a divergent main collimator. A similar *biconical* collimator is also described in /32, 33, 68, 69, 70/ and shown in figs. 22 and 25.

A divergent collimator can be also constructed in the form of *several cylinders with increasing diameter*. Such a construction is described in /53/(fig. 16), /187/ and /95/ (fig. 44).

Several colimators are constructed in *segments,* assembled together, as for example the collimator described in /9, 58, 117/ (fig. 18), /137/ or the conical collimator, described in /135/ which is assembled from three pieces.

In the collimator described in /133/ and shown in fig. 71 the collimator walls are constructed in *stepped* sections, rather than in one continuously diverging form.

Although this report is restricted to the description of divergent collimators it also contains some examples of collimators consisting of a *tube* (/88, 89, 90, 91/ and /120, 121, 122, 123, 124/) or a *cylinder* (/116/ and fig. 64).

Some designs of the collimator consist of a *conical* inner *and cylindrical* outer tubes (e.g. /29/ fig. 9, /85/ or /93/ fig. 43).

The collimation of the neutron beam can sometimes be obtained using *other* means than the collimators described above.

In /95/ a collimating design is described in which 4 collimators are installed within the biological shield of the reactor. A further collimator is situated within a concrete shielding block (fig. 44).

The NR facility described in /102/ uses an arrangement of diaphragms of paraffin and cadmium as a divergent collimator (fig. 52).
Table 2 summarises the various physical shapes of collimators.

Table 2. Shape of the collimators

Shape	Described in reference no:	Shown in fig. no:
Truncated cone	5,19,29,75,85,86,97,107,135,144	3,9,28,37,48,56,78
Truncated pyramid	69,78,79,81	24,32,33
— square	83,84,133	36,71
— retangular	36,49,52,54,55,56,77,86,104,105,106,140	15,17,40,54,55,58,76
Divergent-convergent	69	24,25
Biconical	32,33,68,69,70	22,25
Several cylinders	53,87,95	16,44
Segmented	9,58,117,135	18
Stepped sections	133	71
Tube or cylinder	88,89,90,91,116,120,121,122,123,124	64
Conical + cylindrical	29,85,93	9,37,43
Other	95,102	44,52

5.2. Walls and their lining

Unlike electrons, neutrons cannot be focussed. Therefore the neutron beam must be collimated. To prevent stray neutrons from reaching the radiographed object and to reduce the scattering of neutrons within the collimator the collimator walls must be lined with a material having a high cross absorption section to neutrons. For that purpose boron, cadmium, dysprosium, europium, gadolinium and indium are used. The nuclear effectiveness of these materials can be assessed from fig. 79 /2/. As can be seen the effectiveness of an absorbing material will vary with the neutron-energy spectrum of the neutron beam.

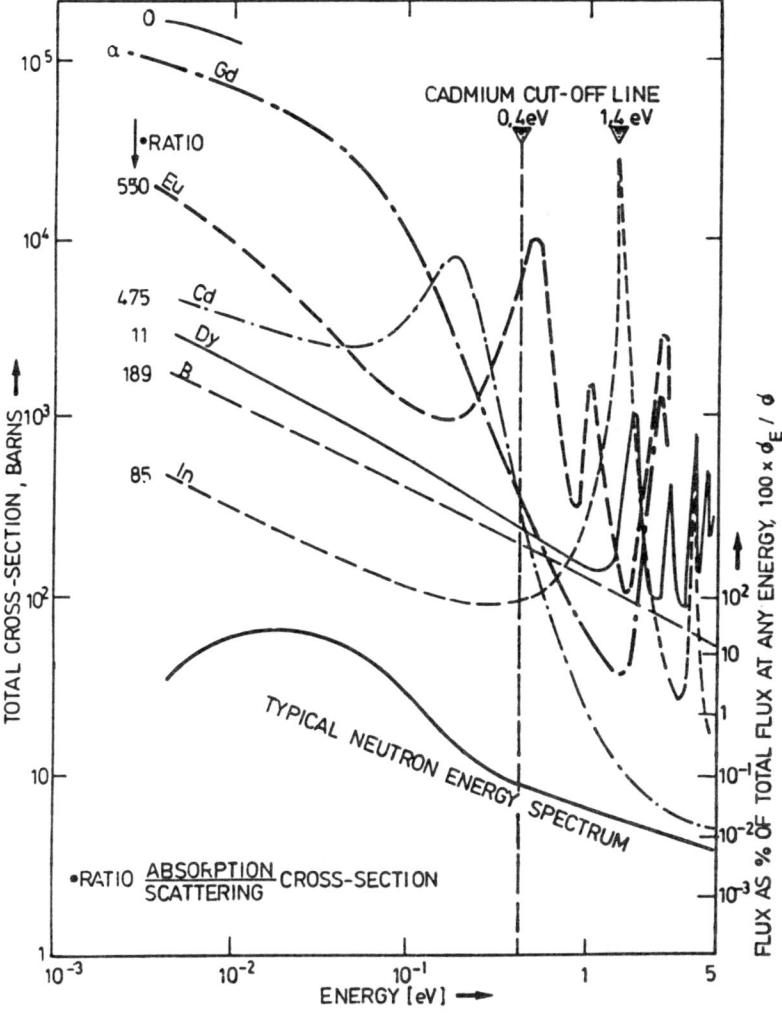

Fig. 79. Total cross-section curves for B, Cd, Dy and In

In /2/ data on boron, cadmium, europium, indium, dysprosium and gadolinium are given (in tables 1.7 to 1.12), where the characteristics of those lining materials are also quoted.

Some practical advice about the use of those materials taken from /2/, has already been quoted in 2 above.

In table 5.1 of /2/ the collimator lining is given for all NR facilities of the European Community.

In most NR facilities the *walls* of the collimator are made of aluminium (sheet or cast) (see table 3).

In several instances stainless steel is used (see table 3).

Table 3. Walls of the collimators

Material	Description in reference no.	Shown in fig. no.
Aluminium	9,29,35,36,49,52,54,55,56, 69,71,75,85,86,88,89,90,91, 107,140	9,15,17,24,28,37, 54,55,56,57,58,76
Stainless steel	5,86,93,112,113	39,43,62

The most important item of each collimator is its *lining*. Here a variety of solutions can be found. The collimator body, made chiefly of aluminium or stainless steel is lined from the inside with materials described below. Only in one instance is a reference found to an outside lining of the collimator /86/. (fig. 39).

A lining with *cadmium* of different thicknesses is described in 26 reports (see table 4).

The use of *cadmium and indium* is described in 3 reports. Also mention is made in two reports of the use of *cadmium, indium and boron* for the lining of the collimator.
A mixture of *lead with 3.5 to 4.0% of cadmium* is used in /135/.
Also a combination of polyethylene and cadmium is used in two NR facilities.

Cast *epoxy resin faced with cadmium* is used in one instance. A lining of the first part of the collimator with B_4C-Cd-In is reported in two instances.

The combination of Cd+Gd+In+Dy+Au lining is used for the whole collimator (fig. 20) or for only the nose of the collimator /77/.

Boral B_4C is equally often used for the lining of the collimator. Its use is quoted in no less than 30 reports. Data about B_4C can be found in /35/ and also some interesting data about the use of boral are given in /53/.

Boral with a combination of a Pb-Gd alloy is used at the inlet of the collimator (fig. 46).

A detailed description of neutron shielding with boral is given in /104/.

The use of B_4C+In is reported in /77/.

A *lead and boron carbide* matrix is selected for the collimator described in /138, 139/.

The collimator described in /141/ is cast from *polyester resin mixed with powdered Pb and B_4C*.

Boron is also used for lining of the collimators. Its use is mentioned in 4 reports, whereas the use of *borated paraffin* is noted in 3 reports.

The use of *lithium* for the lining of the neutron guide is mentioned in /20/.

Collimators made of *lithium fluoride and lead* are described in /95/.

The use of Eu_2O_3-*aluminium* dispersion at the small end of the collimator is noted in /140/.

The use of all the above-mentioned materials for the lining of the collimators is summarized in table 4.

Table 4. Materials used for lining

Material	Thickness (mm)	Described in reference no.	Shown in fig. no.
Cd	0.76	8	1
	1.5	75	28
	0.8	82	32
	3.0	94	
		58,85,88,89,90,91,93	18,37,43,49,
		99,116,130,140	64,69,76
Cd + In		30,35,65	11
Cd + In + B		19,20	3, 4
Pb + Cd		135	72
Polyethylene + Cd	0.8 Cd	81,82	31
Epoxy + Cd		116	64
B_4C + Cd + In		71,86	24,40
Cd + Gd + In + Dy + Au		65,77	20
B_4C		21,22,29	6,9
	15-20	35	
	25	37	12
		49,52	15
	6.35	16	
		54,55,56,71,78,92,96,104,	17,23,30,46,56,57,
		107,111,112,113,117,120,	60,61,62
		121,122,123,124,134,	
	6.35	140	76
B_4C + In		77	
Pb + B_4C		138,139	75
Polyester + Pb + B_4C		141	
B		20,30,35,65	
Borated paraffin		5,95,97	48
Li		20	
LiF + Pb		95	44
Eu_2O_3 + Al	6.35	140	76

5.3. Filling of the collimator

As mentioned in /2/ the attenuation of about 5% per meter in a collimator filled with air can be reduced to less than 1% per meter when helium is used. Therefore in most collimators reviewed in this report *helium* is used /9,20,21,36, 49,52,53,83,84,86,107,111,131,133,134/.

The use of *argon* is also reported in /57,83,84/.

In the in-pool NR facilities compressed gas is used for the filling of the collimator to compensate for the external pressure of the water of the reactor pool.

5.4. Shutters and diaphragms

Although the description of shutters and diaphragms used at the ends of the collimators does not fall directly within the scope of this report, some information is referenced from the literature that are reviewed in this report.

The conclusions about the choice of the best material for the front (inlet) of the collimator, formulated in /2/, are repeated in 2 above. For that purpose a combination of *cadmium and indium* is recommended. Such a solution is adapted for the shutter described in /9/.

A sandwich of Cd and In is chosen to provide the entrance diaphragm in /86/.

In fig. 47 a pinhole arrangement of the collimator at /96/ is reproduced, where a wheel fabricated as a sandwich of 0.75 mm Cd and 1.0 mm In is contained in the heart of the collimator between two 6 mm boral plates.

Due to the high content of epithermal neutrons in /97/ a Cd+In filter is placed at the output of the collimator (fig. 48).

In the NR facility described in /133/ the aperture in the collimator consists of a Cd-Pb-In eutectic mixture, with 0.127 mm thick Gd foil at the 25.4 mm square aperture.

The use of *boral* (B_4C) as the diaphragm material is quoted in several reports. Fig. 53 /2/ shows such a boral diaphragm placed between two Pb shields.

For shutters, /3/ recommends boral sheets, LiF, B_4C+Pb or massive Masonite. Such a shutter made of 12.5 mm thick boral is described in /9/.

In /77/ mobile diaphragms of B_4C+*In* are used.
A B_4C+*Al* diaphragm is once used in /69/.

A shutter of *Pb*+B_4C is reported in /92/.

At the entrance to the collimator, described in /54, 55, 56/, a small Al nose lined inside with *Dy*+*In*+*Gd*+*Cd* (fig. 17) is used.

Attached to the collimator, described in /140/, is a curved plate (with apertures) made of 4.75 mm thick Eu_2O_3-Al dispersion material, encased in 0.79 mm Al.

The description of shutters and diaphragms given above is summarized in table 5.

Table 5. Diaphragms and shutters

Diaphragm (D) or Shutter (S)	Material	Thickness mm	Reference no:	Figure no:
D	Cd, In		2	
D	Cd + In		86	
D	Cd + In	0.75 + 1.0	96	47
Output filter	Cd + In		97	48
D	Cd + Pb + In		133	
	Gd	0.127		
D	B_4C		2	53
D	B_4C + In		77	
D	B_4C + Al		69	
D	Eu_2O_3-Al		140	
Input nose	Dy + In + Gd + Cd		54,55,56	17
S	B_4C		3	
S	LiF		3	
S	B_4C + Pb		3,9	
S	Massive Masonite		3	

5.5. Gamma-ray filters

To filter gamma-rays out of the neutron beam lead or bismuth filters are used. The description of these filters is also outside of the scope of this report. Therefore only a few examples will be selected from the reports reviewed.

The collimator mentioned in /5/ contains an 80 mm thick *lead* filter.
To suppress the gamma-rays a 200 mm thick, ported Pb filter is installed in /8/ (fig. 1).

For the same purpose a lead shield with conical opening is fitted at the entrance of the /53/ collimator.

A Pb plug is inserted at the beginning of the channel in /95/ to reduce gamma radiation.

In one of the NR facilities, described in /9/, a 20 cm *bismuth* plug is used for gamma shielding (fig. 2).

15 and 10 cm thick monocrystal Bi gamma filters are described in /20/. Such a monocrystal Bi filter, 15 cm thick, is also mentioned in /77/. A Bi filter, placed at

the entrance to the collimator /85/ is used, reducing the gamma dose-rate to 2.8 R/h.
A 10 cm thick Bi crystal is used in /94/ at the inlet and outlet collimators (fig. 45).

5.6. Seals for in-pool collimators

Clearly, the collimators used in the in-pool NR facilities must be watertight.

In fig. 9 /29/ an ice seal is shown, preventing water from entering the collimator.

Some details of the watertight joints used in the French NR facilities are described in /30/ and shown in fig. 10.
They consist of elastic, cast with silastene and ice joints.

REFERENCES

/1/ J.P. Barton. Divergent beam collimator for neutron radiography. Materials Evaluation 25, Sept. 1967, 45A

/2/ P. von der Hardt, H. Röttger. (editors) Neutron radiography handbook. D. Reidel Publishing Co., Dordrecht/Boston/London. 1981

/3/ H. Berger. Neutron radiography. Methods, capabilities, and applications. Elsevier Publishing Co., Amsterdam/London/New York. 1965

/4/ M.R. Hawkesworth (technical editor). Radiography with neutrons. Conference held 10-12 September, 1973, at the University of Birmingham. London. British Nuclear Energy Society, 1975.

/5/ N.D. Tyufyakov, A. S. Shtan. Principles of neutron radiography. Atomizdat Publishers. Moscow. 1975

/6/ H. Berger (editor). Practical applications of neutron radiography and gauging. ASTM STP 586. 1976

/7/ J.P. Barton. Neutron radiography – an overview. 5-19 in /6/

/8/ F.R. Swanson, F. R. Kuehne. Neutron radiography with a van de Graaf accelerator for aerospace applications 158-167 in /6/

/9/ K.D. Kok. Neutron radiography of nuclear fuels at the Batelle research reactor. 183-194 in /6/

/10/ Atomic Energy Review, Vol. 15, No 2, June 1977

/11/ M.R. Hawkesworth. Neutron radiography: equipment and methods. 169-220 in /10/

/12/ A.M. Ross. Neutron radiographic inspection of nuclear fuels. 221-247 in /10/

/13/ R.L. Tomlinson. Industrial neutron radiography in the United States of America. 291-326 in /10/

/14/ L.E. Bryant (technical editor), P. McIntire (editor). Non-destructive testing handbook. Second Ed., Vol. 3. Radiography and radiation testing. American Society for Non-destructive Testing. 1985

/15/ H. Berger, D.C. Cutforth, D.A. Garrett, J. Haskins, F. Iddings, R.L. Newacheck. Neutron radiography. Section 12 in /14/

/16/ J.P. Barton. Implementation of neutron radiography. Section 13 in /14/

/17/   La neutronographie. Kodak-Pathé. Division Rayons X. Paris

/18/   La neutronographie. Kodak-Pathé. Division Entreprises et Administartions. Paris. 1974

/19/   C. Desandre-Navarre. La neutronographie-procédé de Controle non destructif. Position de la neutronographie par rapport à la radiographie. 4-13 in /17/

/20/   J.L. Boutaine, G. Breyat, J. P. Perves. Les sources de neutrons utilisées en neutronographie. 14-24 in /17/

/21/   A. Laporte. Installation type auprès d'une pile piscine. 25-28 in /17/

/22/   H. Houëlle, P. Lécorché, D. Morel, H. Revol. Mini-réacteur pour neutronographie. 31-37 in /17/

/23/   H. Berger. Neutron radiography as an inspection technique. Proceedings of the of the Fourth International Conference on Non-Destructive Testing. London. Butterworth. 1964. 113-117

/24/   Proceedings of the Fifth International Conference on Non-destructive Testing, International Conference on NDT. Canada. 1967

/25/   B.L. Blanks, D.A. Garret, R.A. Morris. Improved resolution neutron radiography. 242-247 in /24/

/26/   W.A. Carbinier. Applications of neutron radiography using a pool reactor. 247-250 in /24/

/27/   T. Inouye, S. Kawasaki, N. Wakabayashi, K. Ogawa. The present aspects of neutron radiography in Toshiba. 250-254 in /24/

/28/   Preprints. 6th International Conference on Non-destructive Testing, Session M-Neutronradiography and holography, 1970

/29/   G. Farny. Applications industrielles de la neutronographie. Appareillage et sources. Report No M2. 13-24 in /28/

/30/   J.P. Perves. Neutronographie sous eau. Surveillance de dispositifs d'irradiation actifs. Premier résultats sur la neutrographie avec des neutrons froids. Report No M4. 25-38 in /28/

/31/   F. Lévai. Nondestructive testing by neutron radiography at a small nuclear reactor. Paper B-43. The Seventh International Conference on Non-destructive Testing. Warsaw. 4-8.6.1973.

/32/   A. Laporte, J.P. Boulaumie. Neutron radiography applied to the development of quality control of the pyrotechnic systems of the Ariane launcher. Paper 4K-3. Ninth World Conference on Non-Destructive Testing. Melbourne 1979

/33/ A. Laporte. Controle non destructif par neutrographie auprès du réacteur Triton-applications industrielles. Proceedings of the first European Conference on Non-Destructive Testing. Mainz, 24-26.4.1978

/34/ A. Laporte, G. Bayon. Optimisation d'un "systeme de controle a niveau de qualite constant" – l'installation de neutronographie industrielle "Orphee". Proceedings of the Second European Conference on Non-Destructive Testing. Vienna, 14-16.9.1981. B-10

/35/ C. Desandre-Navarre. La neutrographie sur reacteurs de moyenne puisance. Irradiation facilities for research reactors. Proceedings of the Symposium on Irradiation Facilities for Research Reactors. IAEA-SM-165/44. IAEA Proceeding series STI/PUB/316. 1973

/36/ Nuclear Quality Assurance. Proceedings of a Seminar on nuclear fuel quality assurance. IAEA. Vienna 1976

/37/ N. Maene, H. Tourwe, N. Mostin, E. Pelckmans. Quality control of plutonium oxide fuel elements by neutron radiography. 323-331 in /36/ IAEA-SR-7/20

/38/ S.J. Crutzen, R.P. Debeir, P.S. Jehenson, F. Luchtmans, A. Barasi, T.A. Giorgi, L. Rosai. Use of neutron radiography for quantitative measurements of sorbed hydrogen in getters and quality control of nuclear fuel pins. 333-347 in /36/ IAEA-SR-7/21

/39/ Proceedings of the International Symposium. New Methods of Non-Destructive Testing of Materials and their Application Especially in Nuclear Engineering. Saarbrücken. 17-19.9.1979

/40/ L. Greim, M. Greim, W. Spalthof. Zerstörungsfreie Nachuntersuchung von Brennstoff – und Absorberstäben mit Neutronographie. 263-273 in /39/

/41/ H. Berger, N.P. Lapinski, K.J. Reiman. Neutron laminagraphy of nuclear fuel subassemblies. 275-282 in /39/

/42/ Post-irradiation examination. Proceedings of the Conference held in Grange-over-Sands on 13-16 May 1980. BNES, London, 1981.

/43/ U. Bergenlid, I. Gustafsson. Examination of fuel rods by means of neutron radiography. 45-53 in /42/

/44/ J.P. Barton, P. von der Hardt (editors). Neutron radiography. Proceedings of the First World Conference, San Diego, California, U.S.A., 7-10.12.1981

/45/ J. C. Domanus (editor). International Neutron Radiography Newsletter published in the British Journal of Non-Destructive Testing and Revue Pratique de Control Industriel

/46/ J.P. Barton. Neutron radiography 1964-1977. Distributed by the American Society for Non-destructive Testing. 1977.

/47/ T. Wall, P. Gillespie. Neutron radiography at Lucas Hights. Atomic Energy. July 1975, 7-12

/48/ P.A. Gillespie, T. Wall. Neutron radiography at Lucas Hights. 85-92 in /44/

/49/ Neutron radiography facilities at Mol. Belgium. No 5 in /45/

/50/ H. Tourwe. Neutron radiography at the S.C.K./C.E.N., 107-114 in: Research with BR-2 neutron beams. P. van Assche (editor) BLG-519, 1977

/51/ H. Tourwe. Description and characterization of the BR-1 Neutron radiography facility. 183-189 in /44/

/52/ G. Deprez, N. Mostin. Installation de neutronographie pour controle de crayons combustibles enrichis au plutonium. Nuclear Energy Maturity. Proceedings of the European Nuclear Conference. Paris, 21-25.4.1975. Vol. 7, 79-83. Pergamon Press, New York, Toronto, Sydney, Paris, Frankfurt

/53/ L.Vu Hong. Avant projet du dispositif de neutronographie au Triga. INIS-ml-1882. April 1974

/54/ J.D. Rogers. Neutron radiography in Brazil. 93-98 in /44/

/55/ R. Fuga. Implantacao e desenvolvimento da neutronografia no reator nuclear (IEAR-1) do Instituto de Energia Atomica. IEA-DT-127. Feb. 1979

/56/ R. Fuga. Neutronografia no reator IEA-R1. Informacao IPEN 16. March, 1984.

/57/ J.D. Rogers. Cold neutron spectra of the General Atomic Triga MRF neutron radiography facility. 977-981 in /44/

/58/ A.M. Ross. Neutron radiography at CRNL in AECL-4062. Fall 1971

/59/ McMaster Nuclear Reactor annual research report. McMaster University. Hamilton, Ontario, Canada. January 1978

/60/ Z. Hrdlicka. Development and use of neutron radiography in the Nuclear Research Institute. Radioisotopy. 17 (1) 69-78 (1976)

/61/ K. Cerny, P. Beran. Neutron radiography at the SR-0 reactor. Crack detection 1982. Defektaskopie 1982. Ceskoslovenska Vedeckotechnicka Spolecnost. Prague. Dum Techniky. 1982. 7-11

/62/ J.C. Domanus. Double beam neutron radiography facility of the Research Establishment Risø. Risø-M-1955. September 1977

/63/ Neutron radiography at the Risø National Laboratory. Risø-M-2320, November 1981

/64/ J. Olsen. Risø National Laboratory. Personal communication

/65/ C. Desandre-Navarre, G. Farny, J.P. Perves. La neutronographie sur les réacteurs au C.E.A. CEA-R-4208, 1972

/66/ Industrial neutron radiography for nuclear uses. Presented by C.E.A. Centre d'Etudes Nucleaires de Fortenay-aux-Roses. Section d'Exploatation Triton at Euratom Working Group on Irradiation Devices Harwell meeting, 2-3.7.1974

/67/ M. Faure, A. Laporte. La radiographie avec neutrons auprès de réacteur Triton et ses applications industrielles en France. INOVA 1975. Paris, 9-13.6.1975

/68/ F. Fransetti, A.P. Laporte. Deux réacteurs de recherche-Triton & Nereide au Service de l'industrie. Symposium sur l'avenir des réacteurs de recherche. Grenoble, 16-17.11.1977

/69/ Evolution des moyennes de controle non destructif par neutronographie dans le cadre des activites des Services des Piles de Saclay. 26th plenary meeting of the Irradiation Devices Working Group. Geesthacht, 8-10.10.1980.

/70/ G. Bayon, L. Laporte, J. Le Gal. A review of ten years of operation of the industrial neutron radiography facilities associated with the Triton reactor at the Fontenay-aux-Roses nuclear research center. 67-76 in /44/

/71/ M. Watteau. Examens non destructifs des elements combustibles irradies. Non-destructive testing for reactor core components and pressure vessels. Report of a panel sponsored by IAEA. Vienna, 29.11-3.12.1971. IAEA-145, Vienna 1972, 475-486

72/ G. Farny. Neutron radiography of irradiated fuel elements using cellulose nitrate films. Radiography with neutrons. 115-121 in /4/

/73/ G. Farny. Neutron radiography devices and their own applications. 1st Scientific conference on the peaceful uses of atomic energy for scientific and economic development. Baghdad, 7-11.4.1975. CEA-CONF-3128. FR 7600114

/74/ M. Zacchéo, J. Berthon, G. Uzureau, A. Laporte. Use of a neutron guide in industrial neutron radiography. 99-106 in /44/

/75/ J.P. Barton, J.P. Perves. Underwater neutron radiography with conical collimator. British Journal of Non-Destructive Testing. December 8, 1966., 79-83

/76/ F. Michel. Neutron radiography in Grenoble (I): a complementary means for inspecting nuclear fuels during irradiation. 241-249 in /44/

/77/ Neutronographie. Installations et applications auprès des réacteurs expérimentaux du Centre d'Etudes Nucléaires de Grenoble. PJ/SEREG 910-237/81. 15.4.1981

/78/ M. Houëlle, C. Mercier, H. Reval. Mini-réacteur pour neutronographie (Mirene). Rapport D.S.N. no 26 (S.E.E.S.N.C. no 188). Communication présentée à la conférence BNES /4/

/79/ A. Laporte. Bilan de cinq années de neutronographie industrielle aupres du réacteur Triton et perspectives d'avenir. Paper 3L1 presented at the Eight World Conference on Non-destructive Testing. Cannes, 6-11.9.1976

/80/ Mirene. Neutron radiography minireactor. Commisariat a l'Énergie Atomique 22/79.

/81/ P. Boyer, J. Joannes, H. Houëlle, C. Mercier, H. Reval. Source de neutrons pour neutrographie des combustibles irradies. Development d'un modele pour neutrographie industrielle. Note technique ADAC 71.28. Cadarache, 5.5.1971

/82/ P. Boyer, E. Le Boulbin, P. Millet. Neutron radiography of subassemblies of pins in a hot laboratory. 395-401 in /44/

/83/ Schülken. Neutronenradiographie am FR2. KFK Nachrichten 2, 1973

/84/ H. Schülken. Neutronenradiographie am FR2. KFK 1841. September 1973

/85/ W. Scharenberg, H.J. Bormann. Neutron radiography studies at the FRJ-1 Reactor

/86/ L. Greim, M. Greim, G.W. Schumacher. Neutronenradiographie am Forschungsreaktor Geesthacht und ihre Anwendung bei Bestrahlungsexperimenten. GKSS74/E/37. 1974

/87/ H.R. Ziegert, G. Hüttig, H. Oehler. Neutronenradiographie mit thermischen Neutronen am Rossendorfer Forschungsreaktor

/88/ Y.D. Dande. Neutron radiography. In the annual report of the Nuclear Physics Division. Bhabha Atomic Research Centre Bombay, India 1974. B.A.R.C. - 768, 24-38

/89/ Y.D. Dande, N.C. Jain, R.S. Udyawar. Neutron radiography of ordnance stores, pyrotechnic devices and composite materials. In the Proceeedings of the National Symposium on Industrial Isotope Radiography Bharat Heavy Electricals Ltd., Tiruchirapalli February 26-27, 1976, 372-377

/90/ J.K. Gosh, J.P. Panakkal, K.N. Chandrosekharan, A. Subramonian, P.R. Roy. Penetrating radiation as a tool for quality evaluation of nuclear fuels B.A.R.C. − 1193, 1983

/91/ J.K. Gosh, J.P. Panakkal, P.R. Roy. Monitoring plutonium enrichment in mixed-oxide fuel pellets inside sealed nuclear fuel pins by neutron radiography. NDT International. 16, No 5, October 1983, 275-276

/92/ C.S. Pasupathy, M. Srinivasan, V. Anandkumar, N. Kannan. A ^{233}U fuelled reactor for neutron radiography. 199-207 in /44/

/93/ A. Kasnowo, Djajusman, D. Umar, F.P. Wattimury. Status report on the activity of the neutronradiography group in Pusan Penelitian Teknik Nuklir Bandung. 107-110 in /44/

/94/ Z. Tabatahaian, K.K. Moghadam, N. Mirhabibi. Design parameters of the neutron radiography facility in the Teheran nuclear research center. Sci. Bull. At. Energy Organ. Iran. No.4, 1983, 82-98

/95/ I.Y. Khadduri. A neutron raiography facility on the IRT-2000 reactor. Nucl. Instr. and Methods. 147, 1977, 115-118

/96/ D. Kedem. A method for obtaining a large-area beam from a reactor beam tube for use in neutron radiography. Proceedings of the Symposium on irradiation facilities for research reactors. Teheran, 6-10.11. 1972. IAEA-SM-165/10. IAEA Vienna, 1973

/97/ M. Giannini, G. Pugnetti, M.C. Ramorino, G. Trezza. Neutron radiography facility for inspecting reactor fuel elements at C.S.N. Casaccia. RT/FI (78)17. November 1978

/98/ A. Tsuruno. Inspection of Pu particle in UO_2-PuO_2 pellet by neutron radiography. 365-368 in /44/

/99/ N. Wada, H. Tominga, N. Tachikawa, S. Enomoto, T. Yasui, Y. Yoshida. Neutron and gamma simultaneous radiography using a ^{252}Cf isotopic neutron source. 681-688 in /44/

/100/ E. Hiraoka, M. Fujishiro, Y. Tsujii, J. Furuta, K. Katsuroyama, T. Tsujimoto, K. Yoneda, K. Okamoto. Characteristics of neutron radiography systems using accelerator and a research reactor. 111-118 in /44/

/101/ K. Kanda, K. Yoneda, S. Fujine. Development of an online neutron radiography system of high resolution for nuclear materials. 219-225 in /44/

/102/ J. Altamirano, N. Segovia, M. Monnin. Development of a neutron radiography system at the Triga Mark III reactor of the Nuclear Center of Mexico. 119-126 in /44/

/103/ J.J. Veenema, D. Mesman, H.P. Leeflang. Neutron radiography experiences at the low flux reactor. 127-133 in /44/

/104/ P.J. de Munk, H.P. Leeflang. An apparatus for neutron radiography. RCN-135 March 1971

/105/ J. Bordo, H. P. Leeflang, J.J. Veenema. Neutron radiography installation in the HFR reactor pool. 227-234 in /44/

/106/ Neutron radiography at the HFR Petten. EUR 7915 EN

/107/ H.P. Leeflang, J.J. Veenema. Neutron radiography installation for long fuel rods at the HFR Petten 235-240 in /44/

/108/ Operation of the High Flux Reactor. Programme progress report. January-June & July-December 1984. CEC. JRC Petten

/109/ Operation of the High Flux Reactor. Annual report 1984. EUR 9811 EN

/110/ U. Bergenlid, I. Gustafsson. Examination of fuel rods by means of neutronradiography 329-353 in /44/

/111/ I. Gustafsson, E. Sokolowski. Neutron radiography at the Studsvik R2-O reactor. AE-484

/112/ R.S. Matfield. Neutron-radiography services in Research Reactors Division at Harwell. AERE-R-6372. 1971

/113/ D.J. Taylor. Neutron radiography at Dido reactor. 145-151 in /44/

/114/ A.L. Rogers. G.S. Tuckey. Neutron radiography on the research reactor Herald. 31-38 in /4/

/115/ P.H. White. A.F. Thomas, B.G. Holland, G.S.G. Tuckey. Progress in NR techniques and applications at AWRE Aldermaston. 135-143 in /44/

/116/ M.L. Mullerder, V.J. Hart. Transient neutron radiography on the Viper pulsed reactor. 39-44 in /4/

/117/ L.J. Harrison. Treat neutron radiography facility. 251-256 in /44/

/118/ D.J. Taylor, L.J. Harrison, J.R. White. Neutron radiography of a third-type subassembly. Proc. 23rd Conf. on Remote Systems Technology, 1975, 190-196

/119/ L.J. Harrison, R.M. Conant. R.W. Mouring. Improvement in neutron radiography at Treat. Proc. 25th Conf. on Remote Systems Technology, 1977, 251-258

/120/ W.J. Richards, G.C. McClellan. Hot fuel examination facility neutron radiography reactor design. 257-262 in /44/

/121/ G.C. McClellan, W. J. Richards. Neutron radiography applications and techniques at the fuel examination facility 437-443 in /44/

/122/ W.J. Richards, W.E. Stephens. Neutron radiography facility at the hot fuel examination facility/North neutron radiography facility Proc. 25th Conf. on Remote Systems Technology, 1977, 28-35

/123/ C.H. Cheatle, G.M. Iverson, G.C. McClellan. Handling equipment for the hot fuel examination facility/North neutron radiography facility. Proc. 25th Conf. on Remote Systems Technology, 1977, 36-44

/124/ W.J. Richards, G.C. McClellan. Neutron radiography at the hot fuel examination facility. Proc. 27th Conf. on Remote Systems Technology, 1979, 203-208

/125/ K.G. Golliher. Neutronradiography of Apollo ordnance. 325-332 in /44/

/126/ K.G. Golliher. Neutron radiography feasibility studies for steel examination for the liquid metal fast breeder reactor program. 461-468 in /44/

/127/ K.G. Golliher, L.E. Hanna. Neutron radiography of Apollo ordnance. Meterials Evaluation, 29, No 8, 1971, 165-170

/128/ J.C. Domanus. Travel Report. Second International Conference on Composite Materials and visits to neutron radiography centers in Canada and the USA. Risø-M-1991. July 1978. 120-127

/129/ J.O. Henrie. Atomics International's L-88 nuclear reactor for neutron radiography. Isotopes and Radiation Technology, 9, No 1, Fall 1971, 41-44

/130/ W.L. Whittemore, J.E. Larsen, J.R. Shoptangh. A flexible neutron radiography facility using a Triga reactor source. Materials Evaluation, May 1971, 93-104

/131/ D.E. Schwarzer, G.V. Fitzpatrick. Equipment and techniques for the utilization of neutron radiography with thermionic fuel elements. Gulf-Ga-A- 12521. 1973

/132/ C.E. Leighty. Neutron radiography at the General Electric nuclear test reactor. 153-162 in /44/

/133/ C.N. Janson, Jr., J.P. Barton, E.A. Proudfoot. Neutron radiography at the Hanford Engineering Development Laboratory. Materials Evaluation, June 1979, 55-61

/134/ R.L. Tomlinson, J.B. Henshall. Design of the fuels and materials examination facility (FMEF) neutron radiography facility for irradiated fuel. 271-277 in /44/

/135/ W.J. Richards, R.T. Petersen, J.A. Prindle. The neutron radiography facility at the 3-MW Livermore pool-type reactor. UCRL-51906. 10.09.75

/136/   D.A. Garrett. The macroscopic detection of corrosion in aluminium aircraft structures with thermal neutron beams and film imaging method. NBSIR-78-1434. 7.12.1977

/137/   F.A. Hasenkamp. Neutron radiography facility at Sandia National Laboratories, Albuquerque (SNLA), 163-171 in /44/

/138/   D.M. Alger, S.R. Bull. Development of a high-resolution thermal-neutron radiography facility. Trans. ANS, 14, 1971, 530

/139/   T. Harrison, D.M. Alger, J.R. Vogt. Neutron and X-ray radiography service facilities at the University of Missouri. Trans ANS, 15, 1972, 13-14

/140/   E. Foster, S.D. Snyder, V.A. De Carlo, R.W. Mc Clung. Development and operation of a high-intensity, high-resolution neutron radiography facility. ORNL-4738. December 1971

/141/   J.P. Taft, J.D. Randall. Development and utilization of a neutron radiography facility at the Texas A&M Nuclear Science Center. Trans. ANS 1977, 51-53

/142/   E.R. Kartashev. Neutron radiography facilities using neutron beams from nuclear reactors. 51-57 in /44/

/143/   W.A. Karpeinkin, Ju.D. Karmilitzij, Ju.P. Kormushkin, W.A. Korotaew, W.S. Kuznietzov, G.D. Ljadov, A.N. Maiorov, N.S. Orlov, N.D. Tjufjakov, A.S. Shtan, W.C. Jaskievitz, Ustanovka dla neitronnoi radiografii modeli NR-21R. Radiatsionnaya Tekhnika, 1978, 202-205

/144/   J. Rant, M. Copic, V. Dimic, R. Ilic, M. Najzer, G. Pregl. Neutron radiography at the J. Stefan Institute. 281-290 in /44/